Terahertz Technology

Edited by Borwen You and Ja-Yu Lu

Published in London, United Kingdom

IntechOpen

Supporting open minds since 2005

Terahertz Technology
http://dx.doi.org/10.5772/intechopen.90973
Edited by Borwen You and Ja-Yu Lu

Contributors
Rahul Goyal, Akash Tiwari, Shyamal Mondal, Nisha Flora Boby Edwin, Vaisshale Rathinasamy, Md. Ahasan Habib, Md. Shamim Anower, Md. Nazmul Islam, Rakesh Kaneriya, Gunjan Rastogi, Palash Basu, Rajesh Upadhyay, Apurba Bhattacharya, Rayko Ivanov Stantchev, Emma Pickwell-MacPherson, Dipak Kumar Palit, Rajib Kumar Mitra, Shin'Ichiro Hayashi, Norihiko Sekine, Ilaria Catapano, Sonia Zappia, Lorenzo Crocco, Amin Soltani, Matthias M. Marco Wiecha, Hartmut G. Roskos, Borwen You, Ja-Yu Lu

First published in London, United Kingdom, 2022 by IntechOpen
IntechOpen is the global imprint of INTECHOPEN LIMITED, registered in England and Wales, registration number: 11086078, 5 Princes Gate Court, London, SW7 2QJ, United Kingdom
Printed in Croatia

British Library Cataloguing-in-Publication Data
A catalogue record for this book is available from the British Library

Additional hard and PDF copies can be obtained from orders@intechopen.com

Terahertz Technology
Edited by Borwen You and Ja-Yu Lu
p. cm.
Print ISBN 978-1-83962-612-8
Online ISBN 978-1-83962-684-5
eBook (PDF) ISBN 978-1-83962-685-2

For EU product safety concerns:
IN TECH d.o.o., Prolaz Marije Krucifikse Kozulić 3, 51000 Rijeka, Croatia, info@intechopen.com or visit our website at intechopen.com.

Meet the editors

Borwen You received a BS in Physics from National Changhua University of Education (NCUE), Taiwan in 2000, an MS in Electrical Optical Engineering from National Chiao Tung University, Taiwan in 2002, and a Ph.D. in Photonics from National Cheng Kung University (NCKU) in 2012. He joined the Faculty of Pure and Applied Sciences, Department of Applied Physics, University of Tsukuba, Japan in 2015. Currently, he is an assistant professor in the Department of Physics, NCUE, where he develops advanced terahertz optics based on the core technologies of fiber sensing, artificial material, and system-on-a-chip. He is also a member of the Japan Society of Applied Physics.

Ja-Yu Lu received a BS and MS in Physics from National Cheng Kung University (NCKU), Taiwan in 1998 and 2000, respectively. She obtained a Ph.D. in Photonics and Optoelectronics from National Taiwan University in 2007. In 2008, she joined the Department of Photonics, NCKU, and is currently directing its Terahertz Optics Laboratory as a professor. Her research interests include electromagnetic wave sensing using terahertz (THz) fibers, waveguides, and any integrated THz structure to manipulate THz photons for near-field detection.

Contents

Preface

This book discusses the technology to manipulate terahertz (THz) electromagnetic waves, including imaging, sensing, and optoelectronic studies. THz imaging and sensing technologies are specified based on the technique of coherently pumping and probing THz-electric-field oscillation with a femtosecond laser. Compared to visible and infrared ray technologies, it is not so straightforward to resolve the problems of diffraction limit and special modulation signals. THz signals in imaging and sensing technologies are useful for sensing biological water as well as for inspecting food. Furthermore, microstructures of photonic crystal fibers can both confine THz waves and target samples with the largest overlapping sections and power response. For optoelectronic technology, various models of material polarization waves in solid media are presented in the THz frequency range to approach efficient emission and detection. These optoelectronic models discussed in this book involve nonlinear optic detection, effects of various electrode patterns, high electron mobility transistor devices, and some conductivity theories. The editors sincerely appreciate all the contributing authors for their efforts. This book is a useful resource to motivate readers to join THz technology with relevant fundamentals of optics, photonics, electronics, and electromagnetics.

Borwen You
Assistant Professor,
Department of Physics,
National Changhua University of Education,
Changhua City, Taiwan

Ja-Yu Lu
Professor,
Department of Photonics,
National Chengkung University,
Tainan City, Taiwan

Section 1

Introduction

Chapter 1

Introductory Chapter: Origin of Terahertz Technology

Borwen You and Ja-Yu Lu

1. Photoconductive schemes

Terahertz (THz) technology in history comes from the time-varying Hertzian dipoles based on optical excitation of high-peak-power lasers. The high-peak-power lasers have pulse durations ($\delta\tau$) less than 100 femtoseconds (*fs*) and high coherence within several tens of micrometers (i.e., $\delta x = C\delta\tau$, where δx and C are, respectively, a light speed and a coherent length in free space). The presented THz radiation from time-varying Hertzian dipoles also has high coherence and can be characterized by amplitudes and time- and spatial-dependent wave phases. This *fs*-laser-excited THz radiation thus satisfies the fundamentals of wave optics [1], called THz waves in literatures.

In the pioneering era during or before the 1980s, THz waves were the picosecond (*ps*) electric pulse signals measured from the integrated circuits on silicon-on-sapphire (SOS) substrates [2], but not from air-free space. For general electronic devices, *ps* electric pulses are extremely fast response or switching performance, which results from the rapid capture of optically injected carriers by the large density of material structural defects. Such the time-varying electric pulse currents, or electric oscillation, to dynamic switch one circuit, are presented on the basis of laser illuminance, called a photoconductive performance. The integrated circuits attach the surfaces of photoconductive layers to generate photocurrents for the functions of Hertzian dipoles [2], transmission lines [3, 4], and planar-integrated optoelectronic antennas [5]. For SOS substrates, silicon is an example of the photoconductive layer to be patterned with metal circuits or electrodes. The active spot of the photoconductor surface was one miniature metal gap of an integrated circuit at which *fs* optical pulses are focused [2].

For the presentation of picosecond photoconductive Hertzian dipoles, two identical Hertzian dipoles were used as the transmitting and receiving dipoles for *ps* electric pulses within a 1.1-mm-thick insulating material (alumina) [2]. The configuration is shown in **Figure 1(a)** had two Hertzian dipoles that were oppositely and symmetrically attached to one 1.1-mm-thick alumina slab. The alumina slab was used as the propagation space of *ps* electric pulses. The electric pulse duration within the full width at half maximum was 2.3 ps, corresponding to a coherent length of 0.69 mm and less than the 1.1 mm propagation distance of a dielectric medium (alumina). Compared to the photoconductive electric pulses on the coplanar transmission lines [3, 4, 7] or coaxial cables [8], this is the first demonstration for far-field detecting a free-propagating THz wave without any waveguide support.

Oscilloscopes cannot resolve these *ps* electric pulses in a time domain. The time-varying photocurrent or electric field should be performed from the correlation of electric signals between the pump laser wave and generated THz wave at the receiving Hertzian dipole. **Figure 1(a)** shows the THz waves were synchronously excited

IntechOpen

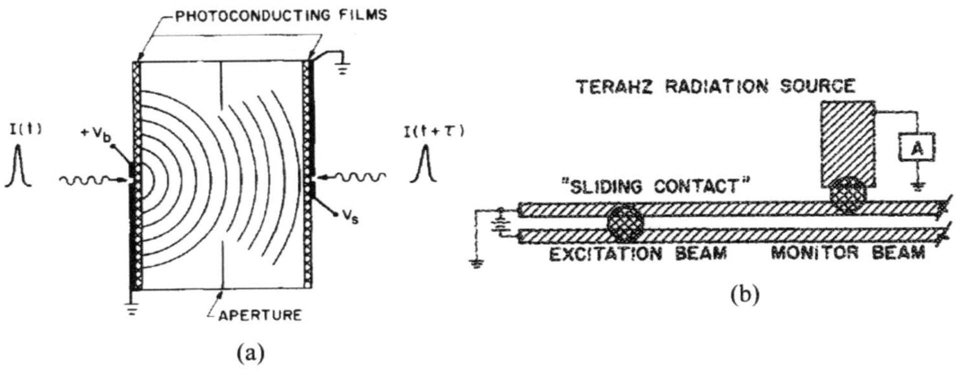

Figure 1.
Configurations to coherently and synchronously generate, detect ps second electric pulses: (a) Hertzian dipoles, (Reprinted from reference [2]. © 1984 AIP publishing) and (b) coplanar transmission lines. (Reprinted from reference [6]. © 1988 AIP publishing).

by one *fs*-pulse laser beam at the transmitting and receiving Hertzian dipoles, respectively, calling excitation and monitor laser beams. A pulse shot of the monitor laser beam [$I(t + \tau)$] at the receiving photoconductor was delayed relative to that of the excitation laser beam at the transmitting photoconductor [$I(t)$], which is controlled by a precise translation stage. The transmitting dipole on the left side of **Figure 1(a)** had a DC bias, and the receiving dipole, which was directly connected to a low-frequency voltage amplifier, detect the bias of the transmitting dipole.

The low-frequency amplifier, i.e., a lock-in voltage amplifier, at the receiving dipole measured the average currents when the optical delay was scanned. The lock-in amplifier integrated one mechanical chopper with a frequency around several KHz. The chopper modulated the pump laser at the transmitting dipole, and the chopping frequency is synchronous to the lock-in amplifier. Based on the signal extraction method, the electric pulse was found down to sub-*ps* level on the coplanar transmission lines, as shown in **Figure 1(b)** [3]. Such the sub-*ps* electric pulse was critically observed when the excitation and monitor laser beam spots are separated as close as possible on the metal circuit due to the lowest substrate dispersion or distortion to the electric pulses.

The point source, THz optics, was first presented from the coplanar transmission lines (**Figure 1(b)**) [6] when one sphere mirror was attached on the sapphire side of an SOS substrate, as shown in **Figure 2(a)** [6]. Different from the electric pulses measured from the alumina-based [2] and transmission-line-based THz devices [3, 4, 7], the electric-pulse waveforms had positive and negative amplitudes with *ps*-scaled pulse widths. It proves that THz wave radiating from one laser spot corresponds one-point source of a sphere wave, whose dimensions are approximate to those of metal gaps but much smaller than those of THz wavelengths, i.e., a subwavelength scale. To efficiently collimate THz waves from the laser excitation on the photoconductive dipoles, high-refractive-index sphere lenses were, therefore, requested to attach to the substrate backsides [6]. The free-space propagation of THz waves was then realized with a high directional feature, like a laser beam, which is shown in **Figure 2(b)** [9]. Based on the photoconductive scheme, THz wave spectrum over 1 THz bandwidth was performed for a 10 cm propagation distance [9]. High-brightness THz waves with a power signal-to-noise ratio (SNR) above 10^5 were then presented when transmission lines and Hertzian dipoles were assembled as the integrated circuits of photoconductive antennas (PCAs), which are shown in **Figure 3(a)** [10]. Furthermore, a suitable optic set up to collect and focus THz waves along the system optic axis was also expressed [10]. **Figure 3(b)** shows the usage

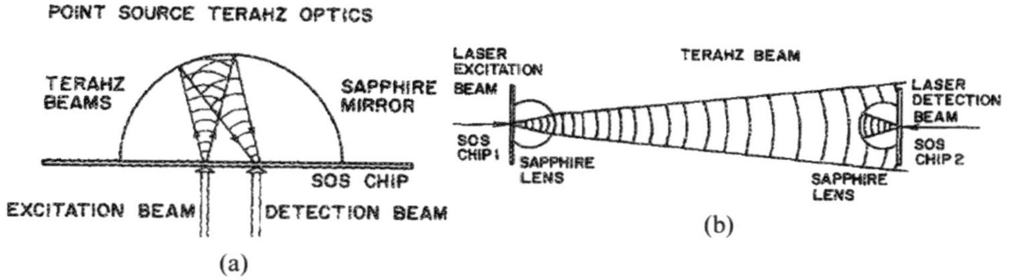

Figure 2.
Configurations of (a) point source THz optics, (Reprinted from reference [6]. © 1988 AIP publishing) and (b) THz beams. (Reprinted from reference [9]. © 1989 AIP publishing).

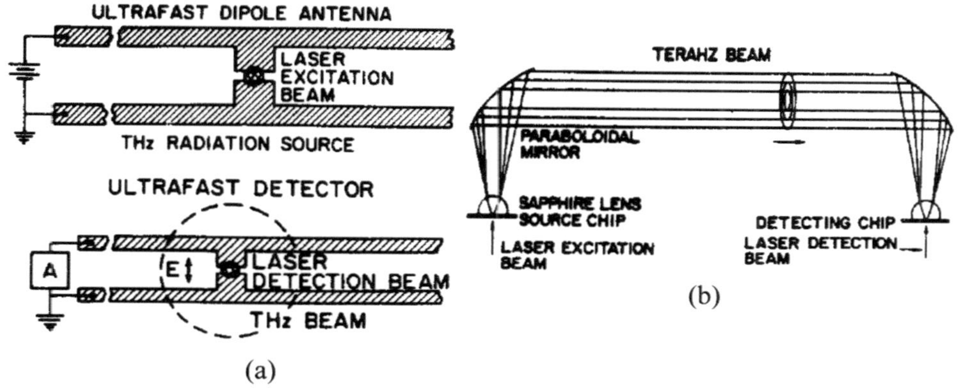

Figure 3.
Configurations of (a) transmission-line-integrated dipole antenna, and (b) off-axis parabolic mirrors between two PCAs. (Reprinted from reference [10]. © 1989 AIP publishing).

of off-axis parabolic mirrors to achieve the mode matching between a THz beam and a *fs* laser beam at a PCA detector [10]. From the development progress of point source THz optics (**Figure 2(a)**), THz beams (**Figure 2(b)**), and the high-brightness schemes (**Figure 3**), the free-propagating THz waves were completely demonstrated and became useful for advanced applications.

2. Bridging THz gaps in optics, photonics, and molecular spectroscopy

In the 1980s, pioneering THz Technology, therefore, opens from the aforementioned THz radiation with the properties of high coherence, high brightness, and high directionality. In 1989, THz spectroscopy based on the photoconductive scheme, called THz time-domain spectroscopy (THz-TDS), was successfully demonstrated to present the molecular absorption lines of ambient water vapor in 0.5–1.5 THz [11]. THz-TDS of water vapor is the milestone of molecular sensing with a fingerprint spectral feature in the THz region. These absorption lines provide the researchers or engineers to calibrate the precision of laser time delay, i.e., exactly extracting THz waveforms in their THz-TDS systems. Using the natural calibrator of THz-TDS—ambient water vapor, THz technology is then extended till the present.

In the 1990s, PCAs were engineered with various integrated circuits [12–14] and various photoconductive layers [15–19] to approach tunable bandwidths and ultrashort pulse widths of THz waves. Besides a photoconductive scheme, nonlinear optics and specified crystals were presented for their fundamentals in THz wave generation [20, 21] and detection [22–29]. The detection limit of PCAs was also discussed in the 1990s [30, 31], and several schemes in the year 2000 were specified to expand the available spectral range of THz-TDS [32–34]. THz imaging concepts were also presented in the 1990s to express more applications of THz waves [35, 36]. In the meantime, the spectral range of THz-TDS, 0.1–3 THz, was considered as one part of the far infrared-ray spectrum to reveal spectral dielectric constants of materials, such as the superconductors [37, 38], toxic chemicals [39], nonpolar/polar liquids [40, 41], vapors/gases [42], and semiconductors (bulks [43], thin film [44] and quantum-well structures [45]). These specified issues lead to the modern THz technology as expressed in this book. For THz-TDS technology, this book introduces the latest progress of THz waves to sense biological water and the possible propagation along photonic crystal fibers. For imaging technology, this book introduces the application of food inspection, the novelty of THz special light modulator, and near-field imaging with a nanometer scale. For PCA technology, the concept of an interdigitated photoconductive antenna is highlighted. For semiconductor technology, the conductivity models are reviewed in this book for the THz field and the relating carrier transition. For nonlinear optics technology, the novelty of THz wave power detection through a harmonic wave conversion and a nonlinear crystal, MgO:$LiNbO_3$, is introduced in this book. These technologies not only bridge THz gaps in optics, photonics, and molecular spectroscopy but also potentially become available in life.

Author details

Borwen You[1,2]* and Ja-Yu Lu[3]

1 Department of Physics, National Changhua University of Education, Changhua City, Taiwan

2 Faculty of Pure and Applied Sciences, Department of Applied Physics, University of Tsukuba, Tsukuba, Japan

3 Department of Photonics, National Cheng Kung University, Tainan, Taiwan

*Address all correspondence to: borwenyou@cc.ncue.edu.tw

References

[1] Saleh BEA, Teich MC. Wave Optics. *Fundamentals of Photonics*. 1st ed. Vol. Chapter 2. Hoboken, NJ, USA: John Wiley and Sons, Inc.; 2001. pp. 41-78

[2] Auston DH, Cheung KP, Smith PR. Picosecond photoconducting Hertzian dipoles. Applied Physics Letters. 1984;**45**:284-286

[3] Ketchen MB, Grischkowsky D, Chen TC, Chi C-C, Duling IN III, Halas NJ, et al. Generation of subpicosecond electrical pulses on coplanar transmission lines. Applied Physics Letters. 1986;**48**:751-753

[4] DeFonzo AP, Lutz CR. Optoelectronic transmission and reception of ultrashort electrical pulses. Applied Physics Letters. 1987;**51**:212-214

[5] DeFonzo AP, Jarwaia M, Lutz C. Transient response of planar integrated optoelectronic antennas. Applied Physics Letters. 1987;**50**:1155-1157

[6] Fattinger C, Grischkowsky D. Point source terahertz optics. Applied Physics Letters. 1988;**53**:1480-1482

[7] Grischkowsky DR, Ketchen MB, Chi C-C, Duling IN, Halas NJ, Halbout J-M, et al. Capacitance free generation and detection of subpicosecond electrical pulses on coplanar transmission lines. IEEE Journal of Quantum Electronics. 1988;**24**:221-225

[8] Mourou G, Stancampiano CV, Antonetti A, Orszag A. Picosecond microwave pulses generated with a subpicosecond laser-driven semiconductor switch. Applied Physics Letters. 1981;**39**:295-296

[9] Fattinger C, Grischkowsky D. Terahertz beams. Applied Physics Letters. 1989;**54**:490-492

[10] van Exter M, Fattinger C, Grischkowsky D. High-brightness terahertz beams characterized with an ultrafast detector. Applied Physics Letters. 1989;**55**:337-339

[11] van Exter M, Fattinger C, Grischkowsky D. Terahertz time-domain spectroscopy of water vapor. Optics Letters. 1989;**14**:1128-1130

[12] Dykaar DR, Greene BI, Federici JF, Levi AFJ, Pfeiffer LN, Kopf RF. Log-periodic antennas for pulsed terahertz radiation. Applied Physics Letters. 1991;**59**:262-264

[13] Darrow JT, Zhang X-C, Auston DH. Power scaling of large-aperture photoconducting antennas. Applied Physics Letters. 1991;**58**:25-27

[14] Hu BB, Froberg N, Mack M, Zhang X-C, Auston DH. Electrically controlled frequency scanning by a photoconducting antenna array. Applied Physics Letters. 1991;**58**:1369-1371

[15] Katzenellenbogen N, Grischkowsky D. Efficient generation of 380 fs pulses of THzz radiation by ultrafast laser pulse excitation of a biased metal-semiconductor interface. Applied Physics Letters. 1991;**58**:222-224

[16] Zhang X-C, Hu BB, Darrow JT, Auston DH. Generation of femtosecond electromagnetic pulses from semiconductor surfaces. Applied Physics Letters. 1990;**56**:1011-1013

[17] Greene BI, Federici JF, Dykaar DR, Levi AFJ, Pfeiffer L. Picosecond pump and probe spectroscopy utilizing freely propagating terahertz radiation. Optics Letters. 1991;**16**:48-49

[18] Xu L, Zhang X-C, Auston DH, Jalali B. Terahertz radiation from large aperture Si p-M diodes. Applied Physics Letters. 1991;**59**:3357-3359

[19] Roskos HG, Nuss MC, Shah J, Leo K, Miller DAB, Fox AM, et al. Coherent

submillimeter-wave emission from charge oscillations in a double-well potential. Physical Review Letters. 1992;**68**:2216-2219

[20] Hu BB, Zhang K-C, Auston DH, Smith PR. Free-space radiation from electro-optic crystals. Applied Physics Letters. 1990;**56**:506-508

[21] Zhang X-C, Jin Y, Ma XF. Coherent measurement of THz optical rectification from electro-optic crystalls. Applied Physics Letters. 1992;**61**:2764-2766

[22] Nahata A, Weling AS, Heinz TF. A wideband coherent terahertz spectroscopy system using optical rectification and electro-optic sampling. Applied Physics Letters. 1996; **69**:2321-2323

[23] Wu Q, Litz M, Zhang X-C. Broadband detection capability of ZnTe electro-optic field detectors. Applied Physics Letters. 1996;**68**:2924-2926

[24] Wu Q, Zhang X-C. 7 terahertz broadband GaP electro-optic sensor. Applied Physics Letters. 1997; **70**:1784-1786

[25] Winnewisser C, Uhd Jepsen P, Schall M, Schyja V, Helm H. Electro-optic detection of THz radiation in LiTaO3, LiNbO3 and ZnTe. Applied Physics Letters. 1997;**70**:3069-3071

[26] Cai Y, Brener I, Lopata J, Wynn J, Pfeiffer L, Stark JB, et al. Coherent terahertz radiation detection: Direct comparison between free-space electro-optic sampling and antenna detection. Applied Physics Letters. 1998;**73**:444-446

[27] Bakker HJ, Cho GC, Kurz H, Wu Q, Zhang X-C. Distortion of terahertz pulses in electro-optic sampling. Journal of the Optical Society of America B: Optical Physics. 1998;**15**:1795-1801

[28] Gallot G, Grischkowsky D. Electro-optic detection of terahertz radiation. Journal of the Optical Society of America B: Optical Physics. 1999;**16**:1204-1212

[29] Gallot G, Zhang J, McGowan RW, Jeon T-I, Grischkowsky D. Measurements of the THz absorption and dispersion of ZnTe and their relevance to the electro-optic detection of THz radiation. Applied Physics Letters. 1999;**74**:3450-3452

[30] Greene BI, Federici JF, Dykaar DR, Jones RR, Bucksbaum PH. Interferometric characterization of 160 fs far-infrared light pulses. Applied Physics Letters. 1991;**59**:893-895

[31] Ralph SE, Grischkowsky D. THz spectroscopy and source characterization by optoelectronic interferometry. Applied Physics Letters. 1992;**60**:1070-1072

[32] Huber R, Brodschelm A, Tauser F, Leitenstorfer A. Generation and field-resolved detection of femtosecond electromagnetic pulses tunable up to 41 THz. Applied Physics Letters. 2000;**76**:3191-3193

[33] Schall M, Uhd Jepsena P. Freeze-out of difference-phonon modes in ZnTe and its application in detection of THz pulses. Applied Physics Letters. 2000;**77**:2801-2803

[34] Kono S, Tani M, Gu P, Sakai K. Detection of up to 20 THz with a low-temperature-grown GaAs photoconductive antenna gated with 15 fs light pulses. Applied Physics Letters. 2000;**77**:4104-4106

[35] Hu BB, Nuss MC. Imaging with terahertz waves. Optics Letters. 1995;**20**:1716-1718

[36] Mittleman DM, Jacobsen RH, Nuss MC. T-Ray Imaging. IEEE Journal of Selected Topics in Quantum Electronics. 1996;**2**:679-692

[37] Nuss MC, Goossen KW, Gordon JP, Mankiewich PM, O'Malley ML,

Bhushan M. Terahertz time-domain measurement of the conductivity and superconducting band gap in niobium. Journal of Applied Physics. 1991;**70**:2238-2241

[38] Gao F, Whitaker JF, Liu Y, Uher C, Platt CE, Klein MV. Terahertz transmission of a $Ba_{1-x}K_xBiO_3$ film probed by coherent time-domain spectroscopy. Physical Review B. 1995;**52**:3607-3613

[39] Flanders BN, Cheville RA, Grischkowsky D, Scherer NF. Pulsed terahertz transmission spectroscopy of liquid $CHCl_3$, CCl_4, and their mixtures. The Journal of Physical Chemistry. 1996;**100**:11824-11835

[40] Pedersen JE, Keiding SR. THz time-domain spectroscopy of nonpolar liquids. IEEE Journal of Quantum Electronics. 1992;**28**:2518-2522

[41] Kindt JT, Schmuttenmaer CA. Far-infrared dielectric properties of polar liquids probed by femtosecond terahertz pulse spectroscopy. The Journal of Physical Chemistry. 1996;**100**:10373-10379

[42] Jacobsen RH, Mittleman DM, Nuss MC. Chemical recognition of gases and gas mixtures with terahertz waves. Optics Letters. 1996;**21**: 2011-2013

[43] Grischkowsky D, Keiding S, Exter MV, Fattinger C. Far-infrared time-domain spectroscopy with terahertz beams of dielectrics and semiconductors. Journal of the Optical Society of America B: Optical Physics. 1990;**7**:2006-2015

[44] Ralph SE, Perkowitz S, Katzenellenbogen N, Grischkowsky D. Terahertz spectroscopy of optically thick multilayered semiconductor structures. Journal of the Optical Society of America B: Optical Physics. 1994;**11**:2528-2532

[45] Waschke C, Roskos HG, Schwedler R, Leo K, Kurz H, Kohler K. Coherent submillimeter-wave emission from Bloch oscillations in a semiconductor superlattice. Physical Review Letters. 1993;**70**:3319-3322

Section 2

Terahertz Imaging and Sensing Technologies

Chapter 2

Terahertz Nano-Imaging with s-SNOM

Matthias M. Wiecha, Amin Soltani and Hartmut G. Roskos

Abstract

Spectroscopy and imaging with terahertz radiation propagating in free space suffer from the poor spatial resolution which is a consequence of the comparatively large wavelength of the radiation (300 μm at 1 THz in vacuum) in combination with the Abbe diffraction limit of focusing. A way to overcome this limitation is the application of near-field techniques. In this chapter, we focus on one of them, scattering-type Scanning Near-field Optical Microscopy (s-SNOM) which – due to its versatility – has come to prominence in recent years. This technique enables a spatial resolution on the sub-100-nm length scale independent of the wavelength. We provide an overview of the state-of-the-art of this imaging and spectroscopy modality, and describe a few selected application examples in more detail.

Keywords: Terahertz, near-field microscopy, s-SNOM, sub-wavelength spatial resolution, THz imaging, Drude conductivity model, plasma waves

1. Introduction

This article deals with THz s-SNOM. The abbreviation stands for Terahertz Scattering-type Scanning-Type Near-Field Optical Microscopy. It is a fairly novel branch of s-SNOM nanoscale spectroscopy which uses terahertz (THz) radiation for the probing (and also excitation) of the interaction with the material under investigation. The THz regime of the electromagnetic spectrum ranges from 0.3 THz to 10 THz, and is situated between the microwave and infrared portions of the spectrum. Interestingly, the THz frequency regime played a significant role during the early development of the s-SNOM technique. Some of the first attempts at reaching spatial resolution much below the diffraction limit were made with THz radiation, at that time using a down-scaled and tapered coaxial line [1]. Inspiration had come at that time from Scanning Near-Field Optical Microscopy (SNOM) which had been invented in 1981 and uses a tapered waveguide, usually a metal-coated optical fiber, with a sub-wavelength diameter at its end, which is brought to within a sub-wavelength distance of the sample. The waveguide is either used for the local excitation of the sample and/or for the collection of radiation from the sample [2, 3]. For the generation of images, the sample is raster-scanned as in all variants of Scanning Probe Microscopy (SPM). This technique works fine with visible and near-infrared radiation, however at longer wavelengths, the throughput of the waveguide decreases drastically if one tries to maintain a spatial resolution as small as in the visible range. In the last decade of the last century, it was then recognized [4], that it is more efficient for near-field sensing with long-wavelength radiation to utilize scattering from the sharp tip of a needle which is illuminated from the outside. As such a needle, it is convenient to employ the probe tip of an Atomic Force Microscope (AFM), as one can then obtain topographical information simultaneously with spectroscopic information. Combined with techniques – described below – to suppress detection of radiation which does not come from the tip (apex) of the needle, this led to the breakthrough of s-SNOM as a powerful near-field spectroscopic technique at the turn of the millennium.

THz radiation always played a role during the rise to prominence of the technique, although a minor one. For example, in 2008, one of the early landmark demonstrations of the usefulness of the s-SNOM technique was the characterization of the spatial profile of the doping density in a Si MOSFET at 2.54 THz (in addition to measurements in the IR) [5]. However, there was always the challenge of the weak scattering efficiency of the probe tip. It arises from the unfavorable ratio of 10^{-4} to 10^{-5} between the probe tip's diameter (typically less than 50 nm) and the diameter of the focal spot of the THz beam (hundreds of micrometers to several millimeters). This challenge contributed to a slow progress which one can discern for THz s-SNOM throughout much of the last two decades, if compared with the huge attention devoted to s-SNOM in the infrared spectral regime. However, the number of publications dealing with THz s-SNOM is rising recently because the THz frequency regime offers a wealth of physical phenomena to be studied.

We mention in passing that also considerable interest exists in the development of s-SNOM-related techniques for the microwave regime. This effort is mainly driven by the need for diagnostic instrumentation for microwave electronic circuits. Again, the first efforts date back to the beginning of the s-SNOM technique itself [6]. Today, this specialized field is known under the name Scanning Near-Field Microwave Microscopy (SMM). A characteristic approach to overcome the challenge of the large wavelength and the concomitant large diffraction-limited focal spot size is the use of the cantilever of the probe tip as a waveguide for the injection and the return of the microwave signal [7]. This is implemented also in commercial products, e.g. from Keysight Technologies. In the following, we will not address the special developments in this field, but only point out the extraordinary attention which is given to careful calibration of the measurement systems which is needed to obtain reliable information about the device-under-test [8].

We also mention briefly other variants of sub-wavelength THz microscopes (both apertureless and with nanoapertures) which are very successfully used in specialized research applications. A preeminent example of an apertureless field measurement technique is the THz sensing of objects-under-investigation which are placed onto an electro-optic (EO) crystal [9]. Here, the THz electric field under the objects is detected by electro-optic conversion in that crystal. The spatial resolution is ultimately determined by the diameter of the focused optical read-out beam, which is in the range of one to several micrometers. An aperture-based near-field technique is the performance of transmission measurements through single slits in thin metallic films [10]. The spatial resolution in one dimension is determined by the width of the slit. With special processes, nanometer-scale widths have been achieved [11]. In order to obtain a sufficiently strong transmittance through the narrow slit to allow for measurements with a good dynamic range, the length of the slit is usually chosen such that a standing-wave resonance occurs at the THz frequency band of interest [12]. With the strong confinement and plasmonic enhancement of the electric field of the radiation in the slit, it has become possible to detect the THz absorption by just a few molecules brought into the slit [13].

Finally, before putting our focus on the s-SNOM technique, we mention two THz near-field techniques which measure electrical currents. Both usually employ THz pulses and time-domain spectroscopy. One technique uses probe tips with integrated photoconductive switches for the detection of electric fields [14]. The other, very novel technique represents an advanced version of Scanning Tunneling Microscopy (STM) and employs THz pulses to induce excitations in the sample which lead to a modification of the tunneling current [15, 16]. This is a very promising technique as it gives for the first time for THz spectroscopy access to the atomic length scale.

2. General aspects of s-SNOM

As stated above, the basis of every s-SNOM device is an Atomic Force Microscope (AFM), where a sharp tip scans the sample to obtain topographical images. s-SNOM now simply adds illumination of the tip-sample region, the radiation inducing a near-field interaction which is detected via the scattered radiation, as shown in **Figure 1**. The red spot indicates the tip-sample region which is illuminated with focused radiation whose electric field ($E_{incident}$) induces a near-field interaction of the tip apex (typical size <50 nm) and the sample. The scattered electric field $E_{scattered}$ contains information about the local dielectric function ε_{sample}. As shown here, one often detects – for practical reasons – the radiation back-scattered onto the beam path of the incoming radiation. The probe tip is in this case part of a cantilever, but recently, researchers also have begun to employ tips attached to a quartz tuning fork, as they are employed in near-field measurement systems which can operate both as AFM and STM systems [17–19]. There are two advantages in the usage of an AFM. One obtains a topographical image simultaneously with the optical near-field information of the sample, and the up-and-down dithering of the probe tip at a frequency Ω provides the basis for the suppression of signals from radiation which does not originate from the apex region of the tip.

Sophisticated interaction models [20–22] and electromagnetic simulations [23, 24] have been published which describe the details of the s-SNOM near-field interaction. The most basic of them, the well-known electrostatic point-dipole model [4, 25] continues to yield valuable qualitative (even semi-quantitative) insight into how the s-SNOM probes the local dielectric function of the sample beneath the tip. This model explains that s-SNOM probes not primarily the surface of the sample but mainly a small volume underneath it. The achieved spatial resolution basically scales with the tip apex geometry, leading to a spatial resolution of typically <50 nm. The resolution is largely independent of the radiation wavelength from optical frequencies [4, 26] down to the sub-THz range [27].

Mathematically, electromagnetic waves under near-field conditions are described as evanescent waves possessing complex wave vectors (k-vectors).

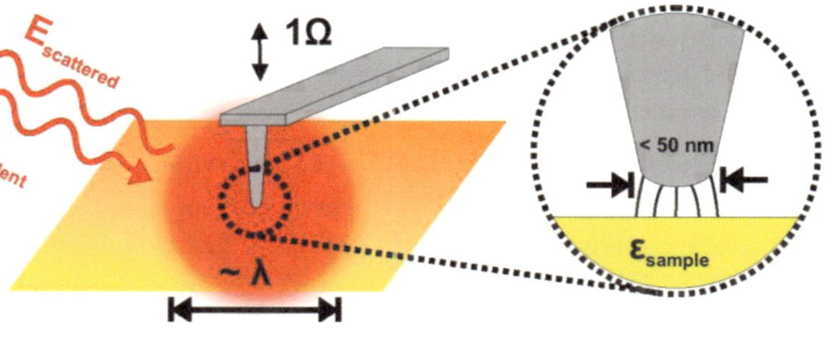

Figure 1.
Scheme of the s-SNOM technique. The AFM cantilever oscillates at a frequency Ω while the surface of the sample is scanned.

Formally, one can understand that sub-wavelength spatial resolution is possible in this case by considering momentum conservation as expressed in Eq. (1), where k_0 is the wave vector of the incoming wave. If one has an imaginary k-component in, e.g., z-direction (k_z), then this allows for large wave vector components in the xy-plane $k_{xy} > k_0$, hence a high spatial resolution in this plane [2, 28]:

$$k_0^2 = k_{xy}^2 + k_z^2 = k_{xy}^2 + (i|k_z|)^2 = k_{xy}^2 - |k_z|^2. \tag{1}$$

The big technical challenge of s-SNOM is already apparent in **Figure 1**. Not only the nanometer-sized tip apex is illuminated by radiation, but rather a spot with an area larger than the diffraction-limited spot size. This spot size scales with λ [2]. Therefore, the back-scattered light contains background light from the sample, tip, and cantilever. This unwanted radiation is much stronger than the desired near-field contribution. As the former cannot be eliminated by wavelength or polarization filtering, it must be suppressed during processing of the detected signal, which is done as follows:

The AFM is operated in the so-called non-contact mode, i.e. the tip oscillates at a frequency Ω close to its mechanical resonance frequency [29]. The different components of the scattered and reflected waves from the tip-sample region are affected in different ways by this oscillation, which allows to cancel their contributions by specific modulation-related filtering measures. A big background contribution arises from (far-field) reflection and scattering from the sample away from the tip apex, resulting in a constant DC term. In contrast, the signal from waves interacting with the oscillating tip away from its apex is modulated at the fundamental oscillation frequency 1 Ω. Finally, the near-field interaction between the tip apex and the sample imprints a higher-harmonic response onto the waves scattered from the apex region. This can be understood in the following way. An essential property of this interaction is its strong nonlinear dependence on the tip-sample distance z (e.g., for the aforementioned point-dipole model, the near-field signal scales with $E_{nf} \propto 1/(1 - bz^{-3})$). Therefore, decomposing the near-field signal from a harmonically oscillating tip ($z = z0 * \sin(\Omega t)$) into its Fourier components leads to a Fourier series of higher harmonics. Hence, demodulation (i.e., extraction) of the measured back-scattered signal at higher harmonics (2 Ω, 3 Ω, … .) isolates the near-field part and suppresses the background [30, 31]. With an increasing order of the higher harmonic, the background suppression becomes better, but the signal strength decreases.

The detectors used for s-SNOM are power detectors. The detected electric field contains the near-field contribution E_{nf} with the phase ϕ_{nf}, and the background contribution E_b with the phase ϕ_b, which yields the total detected intensity

$$I \propto |E_b + E_{nf}|^2 = E_b^2 + E_{nf}^2 + 2|E_b||E_{nf}| \cos\left(\phi_{nf} - \phi_b\right). \tag{2}$$

The higher-harmonic demodulation suppresses the pure background term E_b^2 and, using the approximation $E_b \gg E_{nf}$, the pure near-field term E_{nf}^2 can be neglected as well. Hence, the term $s = 2|E_b||E_{nf}| cos\left(\phi_{nf} - \phi_b\right)$ is the outcome of the higher-harmonic demodulation. As desired, s is proportional to $|E_{nf}|$, but it also depends on the (uncontrollable) background $|E_b|$ and the relative phase between those fields. Hence, a change in the measured signal cannot be attributed unambiguously to the near-field wave, the background or the phase difference between them.

In order to overcome this problem, interferometric frequency-mixing techniques in addition to the higher-order demodulation have been introduced. They play an important role in s-SNOM data recording and processing. Essentially, a second beam of radiation (the reference beam), phase-locked with the beam Different phase-recovering, is projected via a well-defined path onto the detector. The arrangement is interferometric in the sense that it defines a controlled phase-stable situation. The basic idea of the interferometric detection is to make the uncontrollable background E_b insignificant by using the much stronger and well-defined reference field $|E_r| \gg |E_b|$ (whose phase is ϕ_r). The leading contribution to the detected intensity is then given by the signal term $s = |E_r||E_{nf}|\, cos\left(\phi_{nf} - \phi_r\right)$. This brings two invaluable advantages. First, one gains an enormous enhancement of the useful signal by the multiplication of $|E_{nf}|$ with $|E_r|$. Furthermore, with interferometric detection, the optical phase ϕ_{nf} is accessible because ϕ_r is now a known, controlled quantity. With the availability of both the amplitude and phase of the near-field signal, both the real and imaginary part of the dielectric function are accessible. This can be used for the measurement of both the magnitude of the signal attenuation and the phase change by radiation absorption processes [32].

Different phase-recovering detection schemes are used nowadays. Homodyne detection uses a reference beam derived by a beam splitter from the beam sent to the probe tip [33, 34]. Heterodyne detection employs a frequency-shifted reference beam as a local oscillator [26, 35]. Pseudo-heterodyne detection, where the relative phase of the reference beam is varied in the interferometer, is the state of the art of most s-SNOM systems [36, 37]. All those techniques are described and compared in detail in the cited literature.

As the s-SNOM probes the dielectric function $\epsilon(\omega)$, it can be used for the investigation of many kinds of physical phenomena, where a dielectric contrast plays a role. One application area is the distinction of different materials [26]. This can be achieved, for example, by the probing of vibrational fingerprints via IR spectroscopy on a nanometer scale [38]. Another option is the measurement of the electrical conductivity. This was used successfully for the identification of different phases of a material and the observation of phase transitions as a function of time [39]. The sensitivity to local electric fields can also be used for the direct mapping of surface waves (e.g., surface plasmon polaritons [40], phonon polaritons [41], exciton polaritons [42], and more [43]). These polaritons are usually detected as stationary standing-wave interference patterns, but their generation, propagation and decay can also be traced in time with pump/probe techniques. In fact, the first visualization of plasmon polaritons on graphene, excited by the probe tip itself and observed as interference patterns with the returning waves reflected at the boundaries of the graphene flakes [44, 45], has been one of the landmark experiments of s-SNOM.

Although the THz frequency regime benefits enormously from the enhancement of the spatial resolution provided by s-SNOM, for a long time, the well-known "THz gap" (lack of powerful sources and detectors) has been the delaying and limiting factor for the development of a generic THz s-SNOM field of research. This situation is changing only recently.

3. Specific challenges and opportunities of THz s-SNOM

The s-SNOM technique is by now a well-established technique in both the IR spectral range and the visible part of the electromagnetic spectrum [40, 46]. Establishing this technology in the THz frequency regime opens new windows of opportunity because many physical phenomena have characteristic spectral signatures there. In the following section, we describe the state-of-the-art of THz s-SNOM and address specific challenges of s-SNOM at these large wavelengths which had to be overcome or are being addressed presently.

3.1 Time-domain THz s-SNOM

THz Time-Domain Spectroscopy (THz-TDS) is one of the most widespread spectroscopic tools in THz science since it possesses many advantages [47]. It is based on the use of single- or few-cycle THz pulses whose waveforms are measured as a function of time by coherent detection with electro-optic (EO) crystals or photo-conductive antennas (PCAs). With these techniques, one often achieves a high signal-to-noise ratio. Both the amplitude and phase of the electric field of the radiation are recorded. This occurs over the entire spectrum of the pulses which is then available for the extraction of spectral information. Furthermore, high peak intensities are reached, which makes time-resolved pump-probe measurements possible. Its drawback is the poor diffraction-limited spatial resolution which lies in the millimeter regime. The s-SNOM technique can solve this issue and open the path into the nanoworld with sub-100-nm resolution. However, s-SNOM has the following limitation. As the tip oscillates at frequencies of tens to hundreds of kHz, the pulse repetition rate of the THz source must be in the MHz range in order to avoid synchronization challenges with the lock-in detection. Hence, only high-repetition-rate THz sources are employed until now, not amplifier-based systems.

Aligning a time-domain THz s-SNOM system is not trivial since the average beam power is low, on the order of μW, and the THz pulse durations are quite short, typically on the 1-ps time scale. For standard THz-TDS systems, one finds that a high signal-to-noise ratio (SNR) can only be achieved if the measurement system, especially the detection segment, is aligned near its optimum. The alignment of a THz s-SNOM is even more critical because only a small amount of radiation scattered from the probe tip can be detected. As estimated from the data in [17, 48, 49], the signal at the first harmonic 1Ω is reduced to 0.1–1% as compared with the signal obtained in conventional THz-TDS. Concerning the use of THz pulses with ps duration, it is a challenge to find the temporal overlap between the (weak) tip-scattered signal and the laser pulse used for EO detection or for activation of the PCA detector. Upon THz beam steering, the path length changes. In practice, an iterative alignment procedure is employed.

s-SNOM systems operating in the IR, near-IR or VIS preferably exhibit the back-reflection geometry of **Figure 2a**. Here, the scattered and impinging light are on the same side of the probe tip. A beam-splitter (shown in blue in **Figure 2a**) divides the incident light for probing the sample and interferometric detection. The pathway of the incoming radiation – including the focusing leg to the tip –.Time-domain THz s-SNOM systems more commonly use the transmission geometry of **Figure 2b**. The beam alignment is much simplified in this optical arrangement as one can pre-align it with the help of the laser beams used for the generation and detection of the THz pulses. No THz beam-splitter is required. Either a single big paraboloid mirror as in **Figure 2b** or two separate smaller ones are employed for the incoming and the

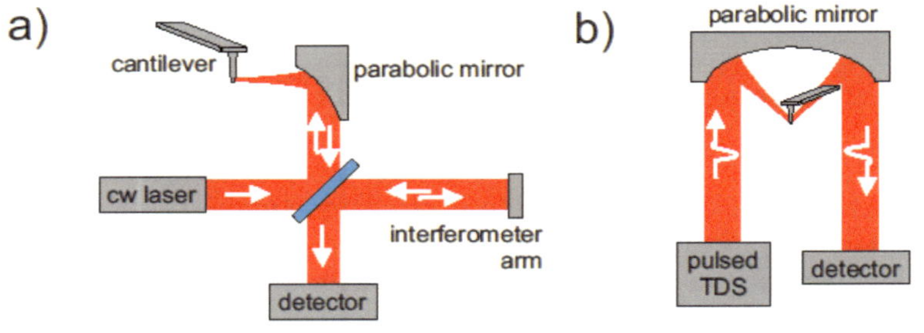

Figure 2.
s-SNOM modalities, (a) back-reflection and (b) transmission geometry.

back-scattered radiation. The coherent detection of the THz pulses, inherent of the THz-TDS approach of **Figure 2b**, eliminates the need for phase-sensitive interferometric detection as it shown in the setup of **Figure 2a**.

In TDS, the measurement of spectra usually requires translation of a mechanical delay stage, which varies the time delay between the scattered radiation pulse and the read-out laser pulse. For the IR region, the term of nano-FTIR (FTIR: Fourier Transform Infrared Spectroscopy) was introduced for this approach [38]. However, the recording of spatial maps across a sample region with full spectral information at each local position is so time-consuming that it is rarely done in the THz frequency range. Nowadays, it is still common to keep the time delay stage fixed at the position of maximal peak intensity during a spatial scan of the sample. This results in maps of the THz response which do not have frequency resolution [49, 50]. Full spectra are often recorded only at selected sample positions [50, 51]. Further improvements are expected because most concepts of time-resolved THz spectroscopy are compatible with s-SNOM, such as Asynchronous Optical Sampling (ASOPS), which does not require a mechanical translation stage for the measurement of spectra [48]. Again the infrared s-SNOM technology is more advanced, as highlighted by Ref. [52], where the propagation of phonon polaritons in hexagonal boron nitride was recorded in real space by time-resolved IR s-SNOM measurements.

3.2 CW THz s-SNOM

A second branch of THz s-SNOM uses continuous-wave (CW) technology. Without the need for finding temporal pulse overlap and with the high spectral power of the radiation, the measurement systems are easier to align than TDS-based systems. The rapid development of this sub-field of THz s-SNOM is reflected in an impressive technological diversity. As THz sources, large-scale facilities such as free-electron lasers (FELs) [53], but also THz gas lasers [5, 33], and compact tabletop electronic (frequency-multiplier-based) sources [27, 35, 54, 55] as well as quantum-cascade lasers (QCL) [19, 56, 57] are being used. The latter is operated in a CW mode, but novel phase-control techniques are currently developed to enable the use of the QCLs in pulsed mode [58]. On the detector side, the selection is limited since the detector must be sensitive and fast enough to resolve the higher harmonics $n\Omega$ of the cantilever frequency in the typical range of several tens to hundreds of kHz, which excludes Golay cells and slow bolometers. In the first years of THz s-SNOM, liquid-helium-cooled hot-electron bolometers were the standard solution [5, 53]. Recently, room-temperature-operated detectors such as Schottky diodes [33, 35, 55] and field-effect transistors [27] have emerged, facilitating the entire measurement procedure. Microbolometers also appear possible, but we are not aware of reports of their use. Finally, THz-QCLs are employed for both emission and detection because the laser light can be generated and measured in a single device [19, 56].

Interferometric detection schemes to achieve more reliable and phase-sensitive measurements become the preferred mode of operation of THz CW s-SNOM. The pseudo-heterodyne scheme as the state-of-the-art technology for s-SNOM in the IR and VIS spectral regimes is challenging to implement for THz radiation. Aiming for the same approach as in the IR/VIS, one would require a mirror oscillating at several hundred Hz with a displacement amplitude scaling with the wavelength [36], which is so demanding that it has not been realized so far. Other approaches appear possible for future use, such as a continuous mirror shift or the use of folded beams. Hence, nowadays, the following alternative interferometric detection schemes find application in CW THz s-SNOM:

- Non-interferometric “self-homodyne” detection is still in use [53], however the phase information is not accessible and this mode possesses the signal ambiguity discussed in conjunction with Eq. (2).

- As the simplest interferometric scheme, homodyne detection is widespread [27, 33]. It offers interferometric signal enhancement and phase sensitivity by two consecutive measurements (at the risk of errors by drifts and tip degeneration). Samples with limited or no optical phase variations across them can be mapped with a single spatial scan by adjusting the interferometer to the maximum signal position.

- Heterodyne detection [35, 55] requires two sources (the second one as local oscillator) and the requirements for the signal-demodulating lock-in amplifiers are high. This makes the approach comparatively cost-intensive, but one benefits from the maturity of heterodyning when working with frequency-stable electronic sources.

- New approaches such as synthetic optical holography (SOH) [59, 60], which relies on data post-processing, have been demonstrated successfully as well.

3.3 Probe tips for THz frequencies: length and shape

THz s-SNOM has also been improved on the basis of a deeper understanding of the fundamental role of the probing tip which acts as an antenna for the impinging radiation. For that reason, metallized tips are preferred; this is the case for IR and VIS radiation, but even more so at THz frequencies. Ref. [61] reports a nine times higher s-SNOM response if standard probe tips, which were developed for the IR, are replaced by longer THz-resonant ones. Nowadays one often finds commercially available Pt tips with a typical length of 80 μm [62] in use, even if they are not exactly resonant at the wavelength of the specific radiation. When self-etched metallic wires mounted on tuning forks [61] are employed, the length of the wires tends to be inadvertently well-suited for THz frequencies.

Another aspect is the tip apex geometry. Surprisingly, blunt tips show a better THz s-SNOM signal than sharp tips [33, 59]. This is explained by the increased interaction volume of the near-field zone to dominate over the reduction of the field enhancement. The loss in spatial resolution due to the bigger tip radius does not scale inversely with the signal gain. Ref. [33] reports a result with one order of magnitude higher THz near-field signal, but inducing a resolution degradation by only a factor of four. Still, one has to make a trade-off between resolution and signal strength.

Regarding the achievable spatial resolution, there seems to be a consensus that routine THz s-SNOM reaches a resolution of 50–100 nm. Best values of <15 nm were achieved in Ref. [33] with customized tips and a well-chosen oscillation amplitude and data averaging. One does not reach the atomic resolution while this is possible with THz-STM [15].

3.4 Ability of THz s-SNOM for depth probing

A new branch of s-SNOM research is dedicated to sub-surface imaging [17] and nano-tomography [63–65] by the analysis of the s-SNOM signals at different higher harmonics (2 Ω, 3 Ω, 4 Ω, … .). They probe the sample down to different depths [63–65]. By the combination of data taken at various harmonics, it is possible to determine depth profiles of specimens.

3.5 Timeline of technological advances of THz s-SNOM

For readers who are interested in a quick overview of important technical developments in THz s-SNOM, the following table presents a timeline of advances with respect to sources, detectors, achieved spatial resolution and other innovations listed under the category "Remarks". The first column specifies the respective reference and the year of its publication. The second column specifies the radiation source; the information "THz pulses" stands for the opto-electronic generation of THz pulses with femtosecond IR or VIS laser pulses. The list does not claim to be complete and only serves as a quick review of the overall development of THz s-SNOM.

4. Application examples of THz s-SNOM

In the following, we discuss four publications in more detail, which are representative for important applications of THz s-SNOM. The topics are the mapping of the conductivity of charge carriers (in Section 4.1), the mapping of surfaces of materials which undergo a phase transition (in Section 4.2), sub-surface imaging (in Section 4.3), and the probing of surface waves (in Section 4.4).

The objects-under-study of all examples are solid-state materials or device microstructures. Extension of THz s-SNOM to probe soft matter is expected in the near future. Similarly, the study of aqueous solutions [66] and biomedical, even alive [67] specimens will come. This will be accompanied with the appearance of modified measurement modalities. Similar developments have taken place in the IR spectral regime where AFM-IR was developed as a family of techniques. The probe tips of AFM-IR do not oscillate, instead acting as contact sensors for measuring thermal expansion induced by wavelength-selective excitation of the specimens [68].

4.1 Nano-imaging of semiconductors with spatially varying density of mobile charge carriers

Determination of the conductivity and of the charge carrier concentration, which is determined by the level of doping, is important for semiconductor technology. THz waves are much more sensitive than IR or VIS radiation for the non-destructive probing of dielectric contrast arising from the Drude conductivity of mobile charge carriers at a density in the range of $10^{16} - 10^{19}$ cm^{-3} (see below), which is the relevant range for semiconductor devices. Ref. [5] employed THz s-SNOM to image this contrast in electronic circuits with field-effect transistors. For that purpose, a processed semiconductor die was cut, and the side facet polished and subjected to s-SNOM imaging. The top panel of **Figure 3a** displays an AFM image. It provides little contrast and the transistor locations cannot be identified. This is possible, however, with the help of **Figure 3b**, which shows an SEM image of the sample. Based on the information from **Figure 3b**, the position of one of the transistors is marked in the top panel of **Figure 3a** by the black box. The two lower panels of **Figure 3a** display the data measured by s-SNOM, namely the 2nd-harmonic-demodulated near-field images recorded at 2.54 THz and 28 THz, respectively. The images exhibit very pronounced signal strength in the regions where one expects doping. The strength of the THz near-field signal is maximal at the device layer and continuously reduces away from the transistors towards the interior of the Si wafer. The decline in signal strength reproduces the gradual decrease of the carrier concentration whose local values (derived from device

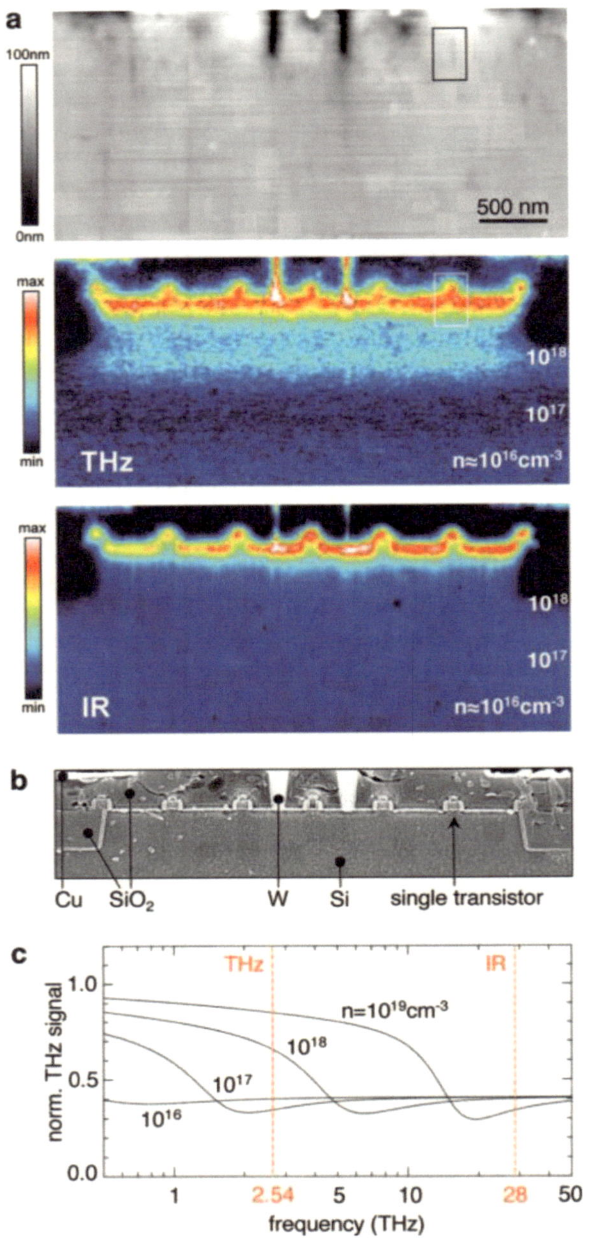

Figure 3.
(a) Three panels showing data measured by AFM (top), 2nd harmonic s-SNOM at 2.54 THz (middle) and 2nd harmonic s-SNOM at 28 THz, respectively. (b) SEM image of the sample surface. (c) Calculated s-SNOM near-field response as a function of charge carrier concentrations. Reprinted with permission from [5]. Copyright (2021) American Chemical Society.

simulations) are listed in the image plots. Compared to the 2.54-THz data, the IR data at 28 THz reproduce the continuous concentration decline only in the vicinity of the devices, but not further away from them. Obviously, IR s-SNOM is sensitive to high carrier concentrations, but not the low ones. **Figure 3c** displays modeled near-field data as a function of the carrier density. It shows the relative signal strength calculated with the point-dipole model for a Drude plasma in Si. The graph confirms that the near-field signal at 2.54 THz is sensitive to doping densities throughout the range specified above. Ref. [5] points out that an average of less than 100 electrons in the probed volume suffices to evoke significant THz contrast. At 28 THz, only the highest densities in the 10^{19}-cm^{-3} range provide a signal sufficiently strong for concentration discrimination. Interestingly, the graph also explains why the 2.54-THz s-SNOM image exhibits a local minimum 900 nm below the transistor layer. The expected signal is minimal at a doping level of 10^{17} cm^{-3} and rises again with decreasing concentration. Finally, we note that THz s-SNOM has recently been used to identify both the carrier type (electrons or holes) and the carrier density [35].

4.2 Phase transitions

Another application of THz s-SNOM is the spatial mapping of phase transitions. Well-known literature examples deal with insulator–metal transitions (IMTs). THz s-SNOM measurements can detect the IMT and allow to decide whether it occurs simultaneously or percolatively across the sample. The measurements also permit to find out the temperature range of the IMT. THz s-SNOM furthermore makes it possible to investigate the formation and thermal evolution of metallic and insulating domains. As an example, Ref. [49] reported measurements of the response of vanadium dioxide at THz and Mid-IR frequencies as a function of temperature, while crossing the transition temperature, T_C, which lies somewhat above room temperature. **Figure 4** shows the demodulated 2nd-harmonic near-field signal measured with a THz TDS system without frequency resolution (top), and the demodulated 3rd-harmonic near-field response in the Mid-IR measured with a CO_2 laser source. The comparison of the results indicates that the signal change at THz frequencies extends over a larger temperature range than in the Mid-IR, where it is fairly abrupt. The Mid-IR data are consistent with a first-order phase transitions, while the THz data suggests a more complex mechanism than the previously assumed one. Besides, coexistence of both insulating and metallic domains are observed at THz and Mid-IR frequencies during the transition.

Another phase-sensing study was reported recently in Ref. [59]. It investigated the chalcogenide $Ge_1Sb_2Te_4$, a so-called phase-change material –a class of materials which is employed for nonvolatile rewritable data storage. Electric or optical pulses can switch these materials between an amorphous and a crystalline phase. In this study, an s-SNOM system with a molecular gas laser and a cryogenically cooled bolometer was used to distinguish the two phases at various frequencies from 1.89 THz to 5.67 THz. The s-SNOM contrast arises from the differences of the phonon spectra of the two phases. The study showed that phase discrimination is principally possible, which opens the way for more detailed studies of domains and of the kinetics of the phase transition.

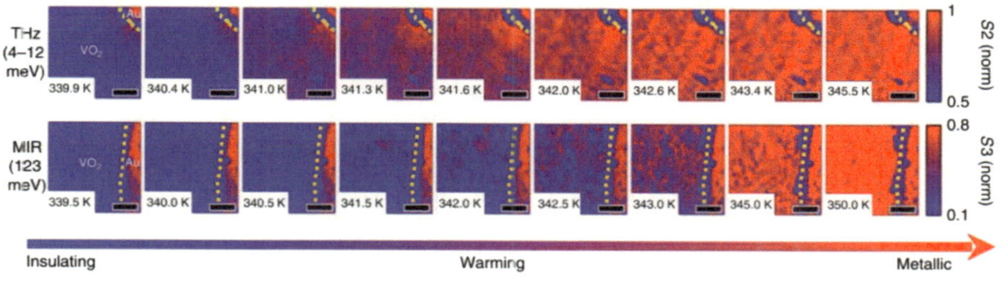

Figure 4.
Near-field mapping of a VO_2 surface during IMT. Measurements were performed as a function of temperature at THz frequencies (top panels) and in the mid-IR (bottom panels). The figure is taken from [49].

4.3 Sub-surface imaging and the potential for nano-tomography

Probing of the depth profile of the permittivity of multilayer samples is discussed in Refs [63–65]. The goal is to unravel or to test the composition of samples with s-SNOM. The publications report measurements and numerical studies for the permittivity of multilayer specimens detected by s-SNOM. The calculations were performed based on full-wave simulations and general solutions of the wave equation by Green's functions. Based on the results, sub-surface tomography aiming at the determination of layer permittivities and thicknesses appears possible within limits if some prior knowledge exists about the composition of the samples. The probing depth is tens of nanometers.

The capability of THz s-SNOM to detect buried structures was reported in Ref. [69]. This represents a simpler task than multilayer tomography. Ref. [67] describes the application of a THz TDS s-SNOM to sense a Au grating embedded in a thin layer of Si_3N_4. The thickness of the nitride layer above the gold stripes varied, but amounted on average to 30 nm. The nitride layer had a smooth surface with the consequence that the topography did not reveal the Au grating, only the permittivity variations did.

Figure 5 depicts the scheme of the sample under the s-SNOM tip that is illuminated by THz pulses. **Figure 6** presents experimental data. Panels (a–c) show the 1st, 2nd and 3rd harmonic of the demodulated near-field signal, respectively. The maps reveal the pattern of the buried Au stripes with a spatial resolution of 90 nm. The contrast arises from the conductivity differences between Au and the insulators Si and Si_3N_4. **Figure 6d** presents the AFM image of the sample, simultaneously recorded with the s-SNOM images. The AFM map delivers no indication for the Au grating. Unlike Ref. [65], no attempt to quantify the depth of the grating based on the s-SNOM data was made. Similar studies were reported in [70], where a buried photonic resonator was studied in the near-IR/VISis spectral regime.

4.4 Direct mapping of surface waves

One of the applications of the s-SNOM technique is the direct observation of surface waves. Recently, Ref. [71] explored the existence of plasma waves in the channels of field-effect transistors (FETs). The excitation of such waves was predicted for the situation that a THz signal is applied either at the gate-drain or the gate-source port of the FET [72, 73]. FETs monolithically integrated with antennas are in use nowadays as detectors of THz radiation. The devices are mainly fabricated in Si CMOS or AlGaN/GaN foundry technology [74, 75]. However, as the

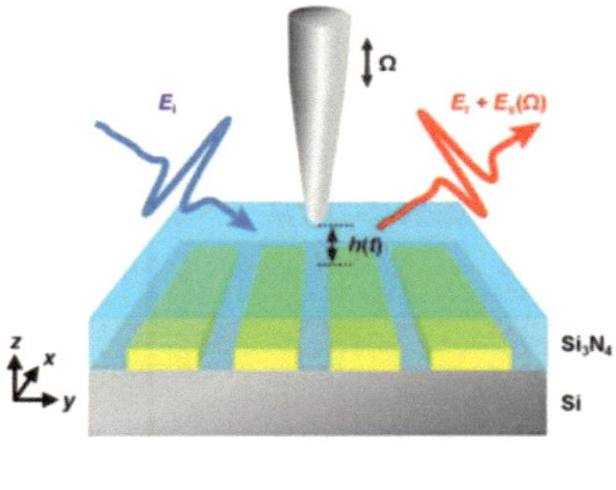

Figure 5.
Schematic of the sample-under-test consisting of periodically arranged Au stripes (thickness: 30 nm, period: 800 nm) embedded in Si_3N_4. The substrate is insulating Si. Reprinted with permission from [15]. Copyright (2021) American Chemical Society.

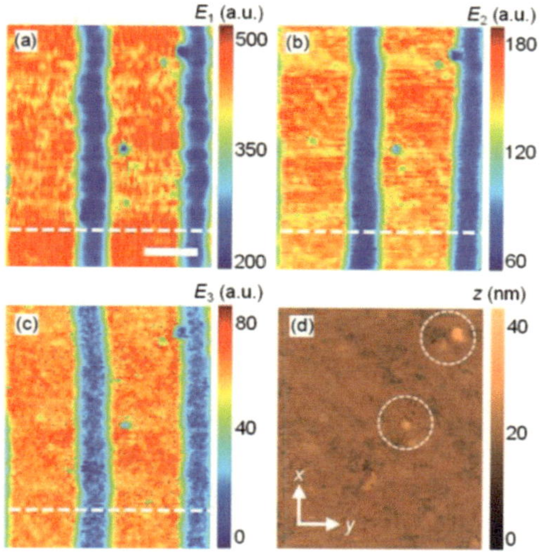

Figure 6.
1st-, 2nd-, and 3rd-harmonic s-SNOM signals (a–c) and an AFM topographical map (d) of the sample depicted in **Figure** *5. Reprinted with permission from [15]. Copyright (2021) American Chemical Society.*

channels are buried under the gate metal, they are not suited for the direct probing of the predicted plasma waves by s-SNOM. Graphene, in contrast, allows for the fabrication of inverted FET structures, where the graphene channel lies on top of the gate.

Figure 7a shows an optical micrograph of a device under investigation in [69]. The graphene FET is located in the center of a THz bow-tie antenna, which is asymmetric for preferred injection of plasma waves from the source side upon THz illumination. **Figure 7b** presents an SEM image of the white box shown in **Figure 7a**. One can discern the graphene stripe on the gate electrode connecting the source and drain electrodes. **Figure 7c** shows a schematic of the device and the s-

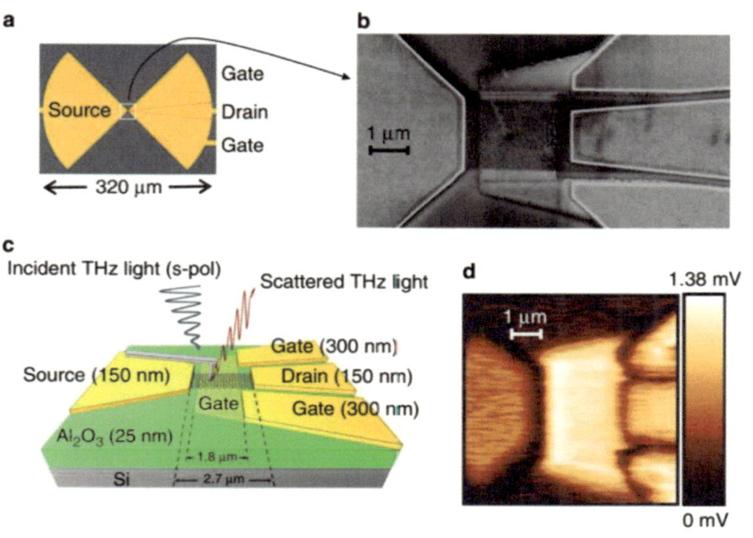

Figure 7.
(a) Optical micrograph of a graphene FET in a bow-tie antenna; (b) SEM image of the FET in the center of the device; (c) schematic of the device and the s-SNOM measurement; (d) near-field map of the electric field distribution derived by signal demodulation at the 2nd harmonic of the tip oscillation frequency. The figure is taken from [71].

SNOM measurement procedure. The device was illuminated by pulsed radiation at 2.0 THz from a FEL, the scattered radiation from the s-SNOM tip was detected with a liquid-helium-cooled InSb hot-electron bolometer. The s-SNOM was operated in self-homodyne mode. **Figure 7d** exhibits a recorded time-averaged electric-field map of the illuminated FET. The bright signal on the gate region has contributions of both the THz signal on the buried gate electrode, which is not fully screened by the graphene sheet, and of the overdamped plasma wave injected into the graphene from the left (source side, see brightest region over the first few hundred nanometers of the graphene stripe). The overdamped plasma wave decays in the graphene sheet over a distance of 0.3–0.5 μm. This is a considerably long propagation distance for the 25–70-fs lifetime of the plasma wave in the unprotected, not encapsulated, graphene sheet.

Another example of THz s-SNOM measurements of surface waves was recently reported in Ref. [76]. Here, the topic was the observation of phonon polaritons in the van der Waals semiconductor α-MoO_3, and the proof of their hyperbolic (self-confined) propagation in spectral regions where the dielectric permittivity is positive in one spatial direction and negative in another. The measurements were performed with pulsed FEL radiation between 8 to 12 THz and cryogenically cooled IR photodetectors. The s-SNOM was operated in self-homodyne mode with 2nd-harmonic demodulation.

5. Conclusion

In this chapter, we have reviewed the state of the art of near-field imaging with the s-SNOM technique at THz frequencies. s-SNOM is an established technique of nano-scale spectroscopy at visible and IR radiation frequencies, and only recently has been extended to longer wavelengths into the THz regime. The same spatial resolution of down to 10 nm is achieved as in the visible and IR spectral regimes. After an introduction of s-SNOM in sections 1 and 2, we have given in Section 3 an overview of the specific technical developments of THz s-SNOM, presenting in **Table 1** a timeline of major advances. Section 4 then has discussed four representative studies which stand for current application areas of THz s-SNOM in the research on solid-state materials and structures. The topics are the mapping of the electric conductivity, the imaging of phase transitions, sub-surface imaging, and the probing of surface waves.

Ref. (year)	Source	Detector	Resolution	Remarks
[77] (2003)	THz pulses	Bolometer EO sampling	150 *nm*	• Stationary tip above the sample
[78] (2007)	THz pulses	Photoconductive antenna	n/a	• Vibrating tip above the sample • Recorded data presented for 1st harmonic
[48] (2008)	THz pulses	EO sampling	n/a	• Self-etched tip • Higher harmonic demodulation
[5] (2008)	THz gas laser	Bolometer	40 *nm*	• Interferometric detection • 2.54 THz
[17] (2014)	THz pulses	Photoconductive antenna	90 *nm*	• Subsurface imaging
[53] (2016)	FEL	Bolometer	50 *nm*	• Broad band • High power • 1.3–8.5 THz

Ref. (year)	Source	Detector	Resolution	Remarks
[61] (2017)	THz gas laser	Graphene-based THz photodetector	n/a	• Tuning tip length in resonance • 3.11 THz
[19] (2017)	THz-QCL	THz-QCL	78 *nm*	• Quantum Cascade Laser self-mixing • Probe tip on a tuning fork • 2.85 THz
[79] (2018)	THz gas laser	Bolometer	~20 *nm*	• Synthetic Optical Holography • 1.89 THz
[35] (2018)	Multiplier chain	Schottky diode (in waveguide)	50 *nm*	• All-electronic THz setup • Transceiver design • 0.5 to 0.75 THz • Heterodyne detection
[33] (2019)	THz gas laser	Schottky diode (in waveguide)	<15 *nm*	• Study the effect of the tip length and shape • 2.52 THz
[54] (2019)	Multiplier chain	Schottky diode (in waveguide)	1μm	• W band (110 GHz) • Large tip instead of commercial AFM tips
[18] (2020)	THz pulses	Photoconductive antenna	<100 *nm*	• Indium tip for a $\frac{\lambda}{2}$ resonance • Probe tip on a tuning fork • Optimized photoconductive antenna for detection
[30] (2020)	Optical and THz pulses	EO sampling	20 *nm*	• THz emission nanoscopy • 11-*nm*laser field confinement
[57] (2020)	THz-QCL	Photocurrent in device-under-test	35 *nm*	• CW THz-QCL as radiation source
[27] (2021)	Multiplier chain	Field-effect transistor (lens-coupled)	40 *nm*	• All-electronic THz setup • FET with monolithically integrated antenna • Phase-resolved homodyne detection • 246.5 GHz

Table 1.
Timeline of advances.

Acknowledgements

M.M. Wiecha acknowledges financial support from the Dr. Hans Messer Stiftung for his doctoral studies. Some of the work was supported by the Hessian Excellence Program LOEWE.

Author details

Matthias M. Wiecha, Amin Soltani* and Hartmut G. Roskos*
Physikalisches Institut, Goethe-Universität Frankfurt, Germany

*Address all correspondence to: soltani@physik.uni-frankfurt.de
and roskos@physik.uni-frankfurt.de

References

[1] Keilmann, F. FIR microscopy. Infrared Phys. Technol. **36**, 217–224 (1995).

[2] Richards, D. & Zayatas, A. *Nano-Optics and Near-Field Optical Microscopy*. (2008).

[3] Pohl, D. W., Denk, W. & Lanz, M. Optical stethoscopy: Image recording with resolution λ/20. Appl. Phys. Lett. **44**, 651–653 (1984).

[4] Keilmann, F. & Hillenbrand, R. Near-field microscopy by elastic light scattering from a tip. Philos. Trans. R. Soc. A Math. Phys. Eng. Sci. **362**, 787–805 (2004).

[5] Huber, A., Keilmann, F., Wittborn, J., Aizpurua, J. & Hillenbrand, R. Terahertz Near-Field Nanoscopy of Mobile Carriers in Single Semiconductor Nanodevices. Nano Lett. **8**, 3766–3770 (2008).

[6] Knoll, B., Keilmann, F., Kramer, A. & Guckenberger, R. Contrast of microwave near-field microscopy. Appl. Phys. Lett. **70**, 2667–2669 (1997).

[7] Tuca, S.-S., Kasper, M., Kienberger, F. & Gramse, G. Interferometer Scanning Microwave Microscopy: Performance Evaluation. IEEE Trans. Nanotechnol. **16**, 991–998 (2017).

[8] Baskakova, A., Stella, G. & Hoffmann, K. An Interferometric Sensor for Scanning Microwave Microscopy Application. in *2018 Asia-Pacific Microwave Conference (APMC)* vols 2018-Novem 1232–1234 (IEEE, 2018).

[9] Seo, M. A. *et al.* Fourier-transform terahertz near-field imaging of one-dimensional slit arrays: mapping of electric-field-, magnetic-field-, and Poynting vectors. Opt. Express **15**, 11781 (2007).

[10] Adam, A. J. L. *et al.* Advanced terahertz electric near-field measurements at sub-wavelength diameter metallic apertures. Opt. Express **16**, 7407 (2008).

[11] Kang, T., Bahk, Y.-M. & Kim, D.-S. Terahertz quantum plasmonics at nanoscales and angstrom scales. Nanophotonics **9**, 435–451 (2020).

[12] Seo, M. A. *et al.* Near field imaging of terahertz focusing onto rectangular apertures. Opt. Express **16**, 20484 (2008).

[13] Park, H.-R. *et al.* Colossal Absorption of Molecules Inside Single Terahertz Nanoantennas. Nano Lett. **13**, 1782–1786 (2013).

[14] Sawallich, S. *et al.* Photoconductive Terahertz Near-Field Detectors for Operation With 1550-nm Pulsed Fiber Lasers. IEEE Trans. Terahertz Sci. Technol. **6**, 365–370 (2016).

[15] Cocker, T. L., Peller, D., Yu, P., Repp, J. & Huber, R. Tracking the ultrafast motion of a single molecule by femtosecond orbital imaging. Nature **539**, 263–267 (2016).

[16] Cocker, T. L. *et al.* An ultrafast terahertz scanning tunnelling microscope. Nat. Photonics 7, 620–625 (2013).

[17] Moon, K. *et al.* Subsurface Nanoimaging by Broadband Terahertz Pulse Near-Field Microscopy. Nano Lett. **15**, 549–552 (2015).

[18] Siday, T., Hale, L. L., Hermans, R. I. & Mitrofanov, O. Resonance-Enhanced Terahertz Nanoscopy Probes. *ACS Photonics* 7, 596–601 (2020).

[19] Degl'Innocenti, R. *et al.* Terahertz Nanoscopy of Plasmonic Resonances with a Quantum Cascade Laser. ACS Photonics **4**, 2150–2157 (2017).

[20] Cvitkovic, A., Ocelic, N. & Hillenbrand, R. Analytical model for quantitative prediction of material contrasts in scattering-type near-field optical microscopy. Opt. Express **15**, 8550 (2007).

[21] McLeod, A. S. *et al.* Model for quantitative tip-enhanced spectroscopy and the extraction of nanoscale-resolved optical constants. Phys. Rev. B - Condens. Matter Mater. Phys. **90**, 085136 (2014).

[22] Esslinger, M. & Vogelgesang, R. Reciprocity Theory of Apertureless Scanning Near-Field Optical Microscopy with Point-Dipole Probes. ACS Nano **6**, 8173–8182 (2012).

[23] Esteban, R., Vogelgesang, R. & Kern, K. Tip-substrate interaction in optical near-field microscopy. Phys. Rev. B - Condens. Matter Mater. Phys. **75**, 1–8 (2007).

[24] Muller, J., Parent, G. & Lacroix, D. Tip optimization for improvement of detection in scanning near-field optical microscopy. J. Opt. **14**, 075703 (2012).

[25] HillenbRand, R., Knoll, B. & Keilmann, F. Pure optical contrast in scattering-type scanning near-field microscopy. J. Microsc. **202**, 77–83 (2001).

[26] Taubner, T., Hillenbrand, R. & Keilmann, F. Performance of visible and mid-infrared scattering-type near-field optical microscopes. J. Microsc. **210**, 311–314 (2003).

[27] Wiecha, M. M. *et al.* Antenna-coupled field-effect transistors as detectors for terahertz near-field microscopy. Nanoscale Adv. **3**, 1717–1724 (2021).

[28] Courjon, D., Bainier, C., Girard, C. & Vigoureux, J. M. Near field optics and light confinement. Ann. Phys. **505**, 149–158 (1993).

[29] García, R. Dynamic atomic force microscopy methods. Surf. Sci. Rep. **47**, 197–301 (2002).

[30] Knoll, B. & Keilmann, F. Enhanced dielectric contrast in scattering-type scanning near-field optical microscopy. Opt. Commun. **182**, 321–328 (2000).

[31] Raschke, M. B. & Lienau, C. Apertureless near-field optical microscopy: Tip–sample coupling in elastic light scattering. Appl. Phys. Lett. **83**, 5089–5091 (2003).

[32] Schmidt, P. *et al.* Nano-imaging of intersubband transitions in van der Waals quantum wells. Nat. Nanotechnol. **13**, 1035–1041 (2018).

[33] Maissen, C., Chen, S., Nikulina, E., Govyadinov, A. & Hillenbrand, R. Probes for Ultrasensitive THz Nanoscopy. ACS Photonics **6**, 1279–1288 (2019).

[34] Gomez, L. *et al.* Apertureless scanning near-field optical microscopy: a comparison between homodyne and heterodyne approaches. J. Opt. Soc. Am. B **23**, 823 (2006).

[35] Liewald, C. *et al.* All-electronic terahertz nanoscopy. Optica **5**, 159 (2018).

[36] Ocelic, N., Huber, A. & Hillenbrand, R. Pseudoheterodyne detection for background-free near-field spectroscopy. Appl. Phys. Lett. **89**, 101124 (2006).

[37] Moreno, C., Alda, J., Kinzel, E. & Boreman, G. Phase imaging and detection in pseudo-heterodyne scattering scanning near-field optical microscopy measurements. Appl. Opt. **56**, 1037 (2017).

[38] Huth, F. *et al.* Nano-FTIR Absorption Spectroscopy of Molecular Fingerprints at 20 nm Spatial Resolution. Nano Lett. **12**, 3973–3978 (2012).

[39] Qazilbash, M. M. *et al.* Infrared spectroscopy and nano-imaging of the insulator-to-metal transition in vanadium dioxide. APS **79**, (2009).

[40] Walla, F. *et al.* Anisotropic excitation of surface plasmon polaritons on a metal film by a scattering-type scanning near-field microscope with a non-rotationally-symmetric probe tip. Nanophotonics **7**, 269–276 (2018).

[41] Dai, S. *et al.* Tunable Phonon Polaritons in Atomically Thin van der Waals Crystals of Boron Nitride. *Science (80-.).* **343**, 1125–1129 (2014).

[42] Hu, F. *et al.* Imaging exciton–polariton transport in MoSe2 waveguides. Nat. Photonics **11**, 356–360 (2017).

[43] Basov, D. N., Fogler, M. M. & Garcia de Abajo, F. J. Polaritons in van der Waals materials. *Science (80-.).* **354**, aag1992–aag1992 (2016).

[44] Fei, Z. *et al.* Gate-tuning of graphene plasmons revealed by infrared nano-imaging. Nature **486**, 82–85 (2012).

[45] Chen, J. *et al.* Optical nano-imaging of gate-tunable graphene plasmons. Nature **487**, 77–81 (2012).

[46] Walla, F. *et al.* Near-Field Observation of Guided-Mode Resonances on a Metasurface via Dielectric Nanosphere Excitation. ACS Photonics **5**, 4238–4243 (2018).

[47] Zouaghi, W. *et al.* Broadband terahertz spectroscopy: principles, fundamental research and potential for industrial applications. Eur. J. Phys. **34**, S179–S199 (2013).

[48] von Ribbeck, H.-G. *et al.* Spectroscopic THz near-field microscope. Opt. Express **16**, 3430 (2008).

[49] Stinson, H. T. *et al.* Imaging the nanoscale phase separation in vanadium dioxide thin films at terahertz frequencies. Nat. Commun. **9**, 3604 (2018).

[50] Yang, Z. *et al.* Near-Field Nanoscopic Terahertz Imaging of Single Proteins. Small **17**, 2005814 (2021).

[51] Aghamiri, N. A. *et al.* Hyperspectral time-domain terahertz nano-imaging. Opt. Express **27**, 24231 (2019).

[52] Yoxall, E. *et al.* Direct observation of ultraslow hyperbolic polariton propagation with negative phase velocity. Nat. Photonics **9**, 674–678 (2015).

[53] Kuschewski, F. *et al.* Narrow-band near-field nanoscopy in the spectral range from 1.3 to 8.5 THz. Appl. Phys. Lett. **108**, 113102 (2016).

[54] Dai, G. *et al.* W-Band Near-Field Microscope. *IEEE Access* **7**, 48060–48067 (2019).

[55] Chen, X. *et al.* THz Near-Field Imaging of Extreme Subwavelength Metal Structures. ACS Photonics **7**, 687–694 (2020).

[56] Giordano, M. C. *et al.* Phase-resolved terahertz self-detection near-field microscopy. Opt. Express **26**, 18423 (2018).

[57] Pogna, E. A. A. *et al.* Unveiling the detection dynamics of semiconductor nanowire photodetectors by terahertz near-field nanoscopy. Light Sci. Appl. **9**, 189 (2020).

[58] Wang, H., Wang, L. & Xu, X. G. Scattering-type scanning near-field optical microscopy with low-repetition-rate pulsed light source through phase-domain sampling. Nat. Commun. 7, 13212 (2016).

[59] Chen, C. *et al.* Terahertz Nanoimaging and Nanospectroscopy of

Chalcogenide Phase-Change Materials. ACS Photonics **7**, 3499–3506 (2020).

[60] Schnell, M., Carney, P. S. & Hillenbrand, R. Synthetic optical holography for rapid nanoimaging. Nat. Commun. **5**, 3499 (2014).

[61] Mastel, S. *et al.* Terahertz Nanofocusing with Cantilevered Terahertz-Resonant Antenna Tips. Nano Lett. **17**, 6526–6533 (2017).

[62] https://rmnano.com/.

[63] Govyadinov, A. A. *et al.* Recovery of Permittivity and Depth from Near-Field Data as a Step toward Infrared Nanotomography. ACS Nano **8**, 6911–6921 (2014).

[64] Mooshammer, F. *et al.* Quantifying Nanoscale Electromagnetic Fields in Near-Field Microscopy by Fourier Demodulation Analysis. ACS Photonics **7**, 344–351 (2020).

[65] Mester, L., Govyadinov, A. A., Chen, S., Goikoetxea, M. & Hillenbrand, R. Subsurface chemical nanoidentification by nano-FTIR spectroscopy. Nat. Commun. **11**, 3359 (2020).

[66] O'Callahan, B. T. *et al.* In Liquid Infrared Scattering Scanning Near-Field Optical Microscopy for Chemical and Biological Nanoimaging. Nano Lett. **20**, 4497–4504 (2020).

[67] Durmaz, Y. C., Goetz, A. & Keilmann, F. Infrared Nanoscopy of Alive Biological Cell Surfaces. in *2019 44th International Conference on Infrared, Millimeter, and Terahertz Waves (IRMMW-THz)* vols 2019-Septe 1–1 (IEEE, 2019).

[68] https://en.wikipedia.org/wiki/AFM-IR.

[69] Moon, K. *et al.* Subsurface nanoimaging by broadband terahertz pulse near-field microscopy. Nano Lett. **89**, 549–552 (2015).

[70] Mey, O. *et al.* Enhancement of the Monolayer Tungsten Disulfide Exciton Photoluminescence with a Two-Dimensional Material/Air/Gallium Phosphide In-Plane Microcavity. ACS Nano **13**, 5259–5267 (2019).

[71] Soltani, A. *et al.* Direct nanoscopic observation of plasma waves in the channel of a graphene field-effect transistor. Light Sci. Appl. **9**, 97 (2020).

[72] Dyakonov, M. & Shur, M. Detection, mixing, and frequency multiplication of terahertz radiation by two-dimensional electronic fluid. IEEE Trans. Electron Devices **43**, 380–387 (1996).

[73] Dyakonov, M. & Shur, M. Shallow water analogy for a ballistic field effect transistor: New mechanism of plasma wave generation by dc current. Phys. Rev. Lett. **71**, 2465–2468 (1993).

[74] Ikamas, K. *et al.* Broadband Terahertz Power Detectors Based on 90-nm Silicon CMOS Transistors With Flat Responsivity Up to 2.2 THz. *IEEE Electron Device Lett.* **39**, 1413–1416 (2018).

[75] Bauer, M. *et al.* A High-Sensitivity AlGaN/GaN HEMT Terahertz Detector With Integrated Broadband Bow-Tie Antenna. IEEE Trans. Terahertz Sci. Technol. **9**, 430–444 (2019).

[76] Oliveira, T. V. A. G. *et al.* Nanoscale-Confined Terahertz Polaritons in a van der Waals Crystal. Adv. Mater. **33**, 2005777 (2021).

[77] Chen, H. T., Kersting, R. & Cho, G. C. Terahertz imaging with nanometer resolution. Appl. Phys. Lett. **83**, 3009–3011 (2003).

[78] Zhan, H. *et al.* The metal-insulator transition in VO2 studied using

terahertz apertureless near-field microscopy. Appl. Phys. Lett. **91**, 162110 (2007).

[79] Mastel, S. *et al.* Understanding the Image Contrast of Material Boundaries in IR Nanoscopy Reaching 5 nm Spatial Resolution. ACS Photonics **5**, 3372–3378 (2018).

[80] Pizzuto, A., Mittleman, D. M. & Klarskov, P. Laser THz emission nanoscopy and THz nanoscopy. Opt. Express **28**, 18778 (2020).

Chapter 3

Spatial Terahertz-Light Modulators for Single-Pixel Cameras

Rayko Ivanov Stantchev and Emma Pickwell-MacPherson

Abstract

Terahertz imaging looks set to become an integral part of future applications from semiconductor quality control to medical diagnosis. This will only become a reality when the technology is sufficiently cheap and capabilities adequate to compete with others. Single-pixel cameras use a spatial light modulator and a detector with no spatial-resolution in their imaging process. The spatial-modulator is key as it imparts a series of encoding masks on the beam and the detector measures the dot product of each mask and the object, thereby allowing computers to recover an image via post-processing. They are inherently slower than parallel-pixel imaging arrays although they are more robust and cheaper, hence are highly applicable to the terahertz regime. This chapter dedicates itself to terahertz single-pixel cameras; their current implementations, future directions and how they compare to other terahertz imaging techniques. We start by outlining the competing imaging techniques, then we discuss the theory behind single-pixel imaging; the main section shows the methods of spatially modulating a terahertz beam; and finally there is a discussion about the future limits of such cameras and the concluding remarks express the authors' vision for the future of single-pixel THz cameras.

Keywords: spatial light modulator, single pixel camera, compressed sensing, Hadamard imaging, terahertz time-domain spectrometer, single-element detectors, sub-wavelength resolution

1. Introduction

Visual information from the eyes generates vast amounts of data for the human brain to process, and provides us with unparalleled clarity and insight into the world we live in. Imaging with terahertz (THz) radiation is a research field that has gained a lot of interest and is in the process of moving from research laboratories to commercial applications [1–4]. As such, the THz research field has grown so much that it has become impossible for a single human to be able to keep track of all developments [4]. Nevertheless, it is possible to outline why there is great interest and potential in THz imaging technology. Most non-conductive materials and non-polar liquids are THz transparent, useful for non-invasive inspection of many multi-component or buried systems, such as paintings [5], electronic circuits [6], space shuttle panels [7] and carbon-fiber composites [8]. Other possibilities are the measurement of picosecond processes in semiconductors [9], quality control of pharmaceutical tablets [10] and non-invasive detection of explosive substances [11]. A plethora of fundamental material resonances, such as phonons, rotations of molecules and precessions of spins, are observable and controllable by THz radiation [12]. Bio-medical applications are highly alluring most notably because the THz photon energies are non-ionizing and high-water sensitivity gives rise to label-free diagnosis of diseases that alter water content, such as cancer [13] and diabetic foot syndrome [14]. There is also the possibility of damaging or repairing DNA with intense THz radiation [15].

With so many possible applications, the reason why THz radiation is barely used outside of laboratories is due to costs of current THz technology. In particular to imaging, the technology is either too expensive, too slow or sacrifices some detection capability (such as picosecond temporal resolution). This is because materials which are suitable for efficient THz detection simply do not exist. This has resulted in THz detector arrays normally working in either narrowbands [16] or needing cryogenic temperatures for sensitive detection [17]. However, microbolometer arrays have very large bandwidths at room temperature operation [18] and when combined with digital holography they can measure both the amplitude and phase of THz radiation [19, 20]. Unfortunately bolometers achieve frequency resolution with a frequency selective source and they do not offer picosecond temporal resolution. This is acceptable for some applications such as detecting concealed weapons, however for applications where time gated detection is used, for example in extracting depths of painting coatings in original art works [5], it becomes unfeasible. An another imaging technique is to project a THz image on an electro-optic crystal then use visible light CCD arrays to spatially map-out the THz field incident onto the crystal [21, 22]. This does not sacrifice the temporal resolution offered time-domain THz spectrometers, however this needs a regen-amplified Ti: Sapphire laser which makes the whole system big and expensive and has prevented the widespread adoption of this technology despite its capabilities. These imaging techniques are all far-field, apart from [21], meaning that they fail to see detail below $\sim 500\mu$m due to the THz wavelengths. There are near-field THz imaging techniques [23–29] which vary from placing AFM-tips next to the sample [23–25], to near-field THz-fiber probes [27] and even air-plasma has been used for sub-wavelength imaging [29]. A very impressive achievement is by M. Eisele, et al. [24] where they obtained 10 nm spatial resolution with 50 fs temporal resolution to reveal the time-dependence of photoexcitation in InAs nanowires.

The aforementioned imaging approaches are the standard imaging techniques, relying on a detector array or raster scanning, however there is another alternative. Namely, using a spatially modulated light beam and a single-pixel detector to obtain an image [30]. Approaches based on this technique are commonly called *computational imaging* because a computer is needed to recover the image although *single-pixel imaging* is another common name. The measurements are obtained sequentially as opposed to in parallel, like in detector arrays, hence this technology is usually slower. This can however be offset by the possibility of obtaining an N-pixel image using fewer measurements by employing techniques known as *compressed sensing* [30, 31]. The main advantage to this technology is that the use of a single-pixel detector greatly enhances the robustness of the system as well as reducing the complexity and cost. This is what makes them attractive for uses in THz-imaging: potential for fast imaging without increasing costs whilst being compatible with single-element detectors such as photoconductive antennas that can measure the amplitude and phase of a THz-pulse with 100 fs temporal resolution. In this chapter single-pixel imaging technologies for use with THz radiation are discussed, where in §2 we discuss the mathematical theory, §3 discusses the current state of the spatial light modulators for THz radiation with regards to the latest implementations of single-pixel THz cameras. §4 discusses where single-pixel THz cameras are likely to have an advantage over other competing THz technologies and the conclusions §5 summarize the whole chapter and ending with the authors' view of the future development of THz single-pixel cameras.

2. Single-pixel imaging theory

Single-pixel imaging theory concerns itself with obtaining an image of a scene using a detector that can only measure the total amplitude emanating from the scene. The simplest idea is to raster scan an aperture across the field-of-view, building the image pixel by pixel. However, as the aperture is made smaller and smaller, the signal reaching our detector is reduced. We could increase the light incident onto our detector and overcome detector-noise by simultaneously scanning more apertures during each measurement, an idea that originates with Yates in 1935 [32]. In **Figure 1(a)** we show the main principle of this idea; we have a light beam that is spatially modulated which propagates through an object and onto a detector with no spatial resolution. It is of the utmost importance that in each measurement we know which apertures were open and which were closed. Without this information we could never reconstruct an image of the object. Each measurement is the dot product of the spatial encoding mask and the transmission function of the object, which is mathematically expressed as

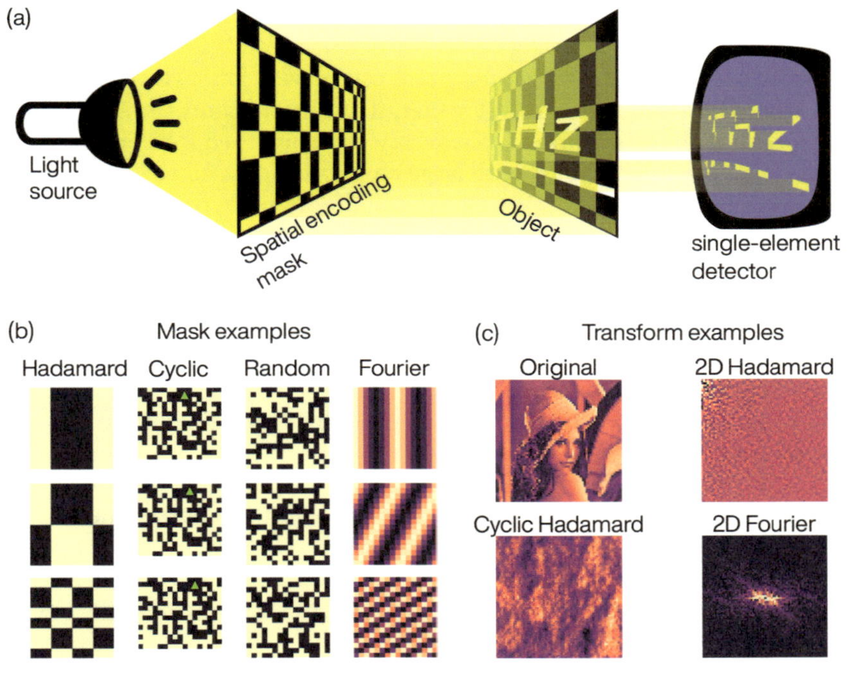

Figure 1.
(a) Imaging with a single-element detector. An encoding mask spatially encodes a beam of radiation, then the beam passes through an object and onto the single-element detector. (b) Spatial encoding masks, where the first, second, third and fourth coloumns were constructed from Sylvester Hadamard, cyclic Hadamard, random and Fourier matrices respectively. The green triangle in the cyclic mask is there as a visual guide. (c) 2D image transform examples. Figure (a) was extracted from reference [33] under creative commons attribution 4.0 international license.

$$y_i = \sum_{j=1}^{N} a_{ij} x_j, \tag{1}$$

where y_i is our i^{th} measurement, a_{ij} holds the spatial information of the ith mask and x_j is the jth pixel of the object. The full set of measurements can be represented by the matrix equation

$$\mathbf{y} = A\mathbf{x}, \tag{2}$$

where the rows of matrix A are shaped into the projected masks (ie. 1D vectors reshaped into 2D masks). Solving Eq. (2) for $\mathbf{x}$ is the cornerstone of single-pixel imaging. In the simpler cases A is an invertible matrix, meaning that $\mathbf{x} = A^{-1}\mathbf{y}$ is sufficient for us to the obtain our image. Alternatively we need to use least squares solvers or complex optimization problems [31], which are commonly used for undersampling procedures where one obtains an image with N pixels using fewer measurements.

A in Eq. (2) is called the basis expansion of our object $\mathbf{x}$. For example, the most commonly know example is if A is the Fourier matrix, then $\mathbf{y}$ is the Fourier transform of our object $\mathbf{x}$. This basis expansion determines the masks that will be projected in each measurement, and in 1(b) we show commonly used spatial encoding masks. Namely, masks constructed from a Sylvester-Hadamard, Cyclic-Hadamard, Random and Fourier matrices. Each one has its own advantages and disadvantages and these are briefly outlined.

The Sylvester-Hadamard matrices are binary, meaning that they are easily implemented, specifically with digital micromirror devices which are relatively cheap and have switch rates upto 20 kHz. The reconstruction technique can also be efficiently calculated by just doing the Fast Hadamard-Walsh transform, meaning one does not need to store A in memory that can be very large (ie. for an $n \times n$ image A would be a $n^2 \times n^2$ matrix). Further, if one implements the 2D transform of the Hadamard-Walsh transform with *sequency* ordering, then it has the effect of concentrating all the coefficients with large amplitudes around the (0, 0) pixel as can be seen in **Figure 1(c)**. This is useful for undersampling procedures with the relevant masks obtained by transforming a series 2D-delta functions [34]. Finally, these are orthogonal matrices meaning they are very robust to detector noise [35]. A side note is that creating such masks at sub-THz wavelength resolutions can create grating diffraction effects as shown in § 5.3.2 of reference [33].

The Cyclic-Hadamard matrices, also known as Paley Type I and type II Hadamard matrices as they were first discovered by Paley in 1933 [36], are orthogonal and circulant matrices made of 1 s and -1 s. This means they have large noise robustness, easy binary implementation and they are constructed by having one vector, $\mathbf{c}$, from which every row is created by cyclic permutations of this vector. For the spatial masks, this means that $i+1$ mask is obtained by shifting the i mask to left which can be seen in column two of **Figure 1(b)**. This means that these masks are most commonly used as physically manufactured masks and can be placed on spinning discs. These matrices also have a fast-reconstruction technique which based on their circulant nature. Given an equation $\mathbf{y} = C\mathbf{x}$ where is a C is matrix constructed by cyclically shifting the top row $\mathbf{c}$, then we can rewrite it as $\mathbf{y} = \mathbf{c} \circledast \mathbf{x}$ with $\circledast$ being the circular convolution operator. Then by the circular convolution theorem $\mathcal{F}[\mathbf{y}] = \mathcal{F}[\mathbf{c} \circledast \mathbf{x}] = \mathcal{F}[\mathbf{c}]\mathcal{F}[\mathbf{x}]$, where $\mathcal{F}$ is the Fourier transform. This means that $\mathbf{x} = \mathcal{F}^{-1}\left[\mathcal{F}[\mathbf{y}]/\mathcal{F}[\mathbf{c}]\right]$, thus if one uses the Fast Fourier Transform then the reconstruction algorithm is very efficient with a complexity of $O(N \log N)$ for an N-pixel image.

Random masks constructed from Bernoulli matrices, or Gaussian random matrices, can also be made from 1 s and -1 s making for easy implementation using binary spatial light modulators. However, these matrices are not directly invertible and using a pseudo-inverse can create stability problems. Therefore convex minimization algorithms are usually used for image reconstruction [30, 31]. The main benefit of this masking approach is that is can be used for undersampling which can greatly reduce the total measurement time at the expense of complicated calculations. References [37, 38] were the first theoretical investigation and one can obtain their reconstruction scripts from reference [39], although reference [40] also freely provides their MATLAB scripts for another minimization algorithm called TVAL3 [41]. These algorithms can be slow, hence a mention needs to be given to reference [42] where Kowarlz et al. creates a pseudo-inverse matrix via Fourier-domain regularization that is able to recover images of quality similar to the slow minimazation algorithms, however with faster calculations based on matrix multiplication methods. Note, they also provide their MATLAB and Python scripts freely on github [43].

Fourier masks are those derived from the Fourier matrix. However, as this is just linear algebra representation of the Fourier transform and we are measuring real images (without imaginary numbers), then we do not need to measure the negative Fourier frequencies as they are just the complex conjugate of their positive frequency counterpart. The Fourier matrix is also orthogonal meaning it has noise robustness equal to the Hadamard matrices as well efficient image reconstruction algorithms, simply the Fast Fourier Transform. These masks, however, are not binary but require grayscale values which limits their deployability. Binary spatial modulators can accomplish this either by temporal dithering, at the expense of slower switch-rates, or by spatial dithering, which creates some quantization errors [34]. Nevertheless, these masks benefit from extensive literature based on the Fourier Transform and various image compression algorithms that can be reversed for image-undersampling procedures.

3. Spatially modulating THz radiation

In this single-pixel imaging modality, the most crucial part is to create a spatially modulated beam. In this respect for the THz regime there are four main methods that can be employed; by creating a physical mechanical mask, by changing the electrical conductivity of a material via the injection/depletion of charge carriers, by controlling the refractive index of liquid crystal cells and by creating a spatially varied beam directly at the THz generation stage.

3.1 Mechanical masks

Creating a physical mask to modulate THz radiation has the great advantage that this is the easiest in terms of manufacturing with great modulation depth, > 99%, over very broadband frequency ranges. This is because most semiconductors and plastics are THz transparent whereas conductive metals absorb and reflect THz radiation [44]. Thus PCB manufacturing techniques can be used to create a set of spatial masks. This results in masks that need to be mechanically moved, however if placed on disc high switch-rates could be potentially achieved as discs can spin at high speeds. These techniques were used in references [45, 46].

There is another modulation technique that falls in this mechanical category. Namely, mirror arrays where each mirror can be individually addressed. Such arrays already exist for the visible light regime in the form of digital micromirror arrays (DMD). However, DMD mirrors are with dimensions around 10 μ m meaning their THz diffraction efficiency is incredibly low. Therefore larger mirrors were manufactured by reference [47]. This means they have good working efficiency and high contrast. **Figure 2** shows their concept in parts (a-c) and their results in parts (d-f). The main problem with these devices is the mirrors have to physically move which creates some limits in regards to the switch rate, otherwise the metallic nature of the mirrors means they have a very broadband modulation frequencies.

3.2 THz modulation by charge carrier injection/depletion

The main principle with this THz modulation technique is based upon the Drude model dielectric function [48, 49].

$$\varepsilon(\omega) = \varepsilon_{\infty} - \frac{\omega_p^2}{\omega^2 + i\omega/\tau_s}, \tag{3}$$

where τ_s is the carrier scattering time, ε_{∞} is frequency independent permittivity due to bound charges, and $\omega_p = \sqrt{Ne^2/m_e\varepsilon_0}$ is the plasma frequency of the material where N is the density of charge carriers with charge e, ε_0 is the vacuum permittivity and m_e is the effective mass. When the incident wave frequency is higher than the

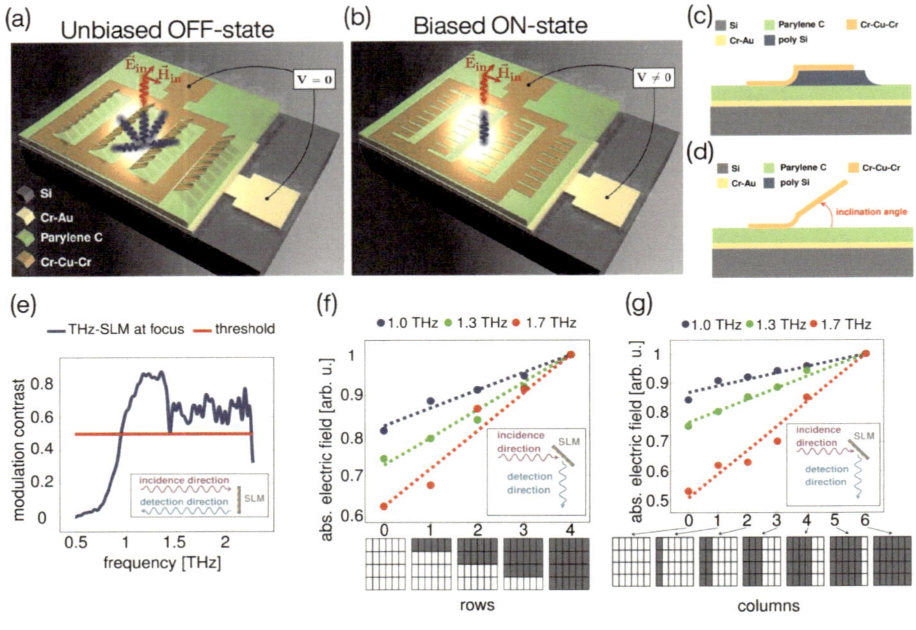

Figure 2.
Schematic of a single pixel of the THz-SLM for normally incident terahertz waves. The pixel is composed of mirrors that are arranged in 4 rows and 8 columns. (a) OFF-state for a bias voltage of 0 V. all mirrors are inclined and incident terahertz radiation (red) is diffracted away (blue) from the transceiver. (b) ON-state for a bias voltage of 37 V. all mirrors are pulled down to the substrate and incident terahertz radiation (red) is reflected (blue) into the transceiver. (c) Schematic cross-sectional view of an unreleased mirror. The base of the mirror adheres to the parylene C, while the part to be released sits on the poly-Si. (d) Schematic cross-sectional view of a released mirror. The base of the mirror adheres to the parylene C, while the released part is inclined due to residual stress in the Cr-Cu-Cr mirror material. (e) Modulation contrast of the THz-SLM. The contrast exceeds a value of 0.5 for a working range from 0.97 THz to 2.28 THz with a maximum contrast of 0.87 at 1.38 THz. (f) Linear dependence of the detected modulated electric field on the number of switched-ON rows in the THz-SLM. (g) Linear dependence of the detected modulated electric field on the number of switched-ON columns in the THz-SLM. Figure reprinted from reference [47] under creative commons attribution 4.0 international license, with figures being relabelled and re-scaled.

plasma frequency, $\omega > \omega_p$, the material has a dielectric response whereas for $\omega < \omega_p$ we have a metallic absorbing material response. Therefore, by controlling N we can change the plasma frequency of the material so the THz wave interacts with a dialectic or a metallic object. This is the main modulation principle of these devices. However, vanadium dioxide (VO_2) needs to be mentioned as it can undergo an insulator-to-metal phase transition via femtosecond optical excitation [50] or an electric field [51]. This material is placed here as using it as a THz modulator has identical experimental implementation, despite the fundamental physics being very different; namely THz modulation via electrical gating or optical excitation.

3.2.1 Optical modulators

Optical based spatial THz-light modulators are currently the best in terms of achieved switch-rate, operational frequency and ease of implementation. Their switch-rate and operational frequencies are both similar to the electrical modulators in that they also rely on modifying the charge carrier density in some material. However, as they use optical light to achieve this, their experimental implementation is very different and due to the current state of visible-light SLMs they are much easier to be implemented. One starts by patterning a visible light beam and then projecting this spatial pattern onto a semiconductor, thereby creating areas that experience large optical excitation and other areas which are left in their ground state. This in turn creates a spatially varying conductivity/absorption profile on the surface of the semiconductor, and thus if a THz beam passes through this surface then the inverse spatial pattern from the visible-light beam is imparted onto the THz beam. The optical excitation can come in two forms, pulsed and continuous wave. For both cases, the carrier concentration is described by

$$N(t) = e^{-t/\tau_l} \int_{-\infty}^{t} G(t') e^{t'/\tau_l} \mathrm{d}t' \tag{4}$$

for carrier generation rate $G(t)$ and excited carrier lifetime τ_l. Reference [52] developed a post-processing technique to account for the above equation, resulting in significantly improving image signal-to-noise when the masks are switched on time-scales comparable to τ_l. However, it should be noted that their mathematical proposal is more general and is able to account for any other physical processes, relevant to the temporal response of the system, that can be described by mathematics.

For continuous wave excitation one needs to consider the photo-carrier generation, recombination and diffusion dynamics within the semiconductor. The steady-state equilibrium carrier concentration within the semiconductor is given by [48].

$$N = \frac{\text{Carrier generation}}{\text{Carrierre combination}} = \frac{2 I_0 T \tau_l}{V \hbar \omega_e}, \tag{5}$$

where I_0 is the incident power per unit area, T is the Fresnel transmittance of the semiconductor, V is the volume where the photocarriers are in, $\hbar\omega_e$ is the photon energy of the photo-excitation beam, and the factor of 2 accounts for the excitation of electrons and holes[1]. The modulation depth is going to be the difference between the ground-state carrier concentration and this new photomodulated concentration. In Eq. (5) it can be seen that large τ_l will give the greatest modulation depth, however this comes at the expense of switch-rate. This is seen in **Figure 3(a)**, where

---

[1] This assumes that the photogenerated holes and electrons have similar properties such carrier mobility, effective mass, lifetime,... ie. a hole and an electron absorb the same amount of terahertz.

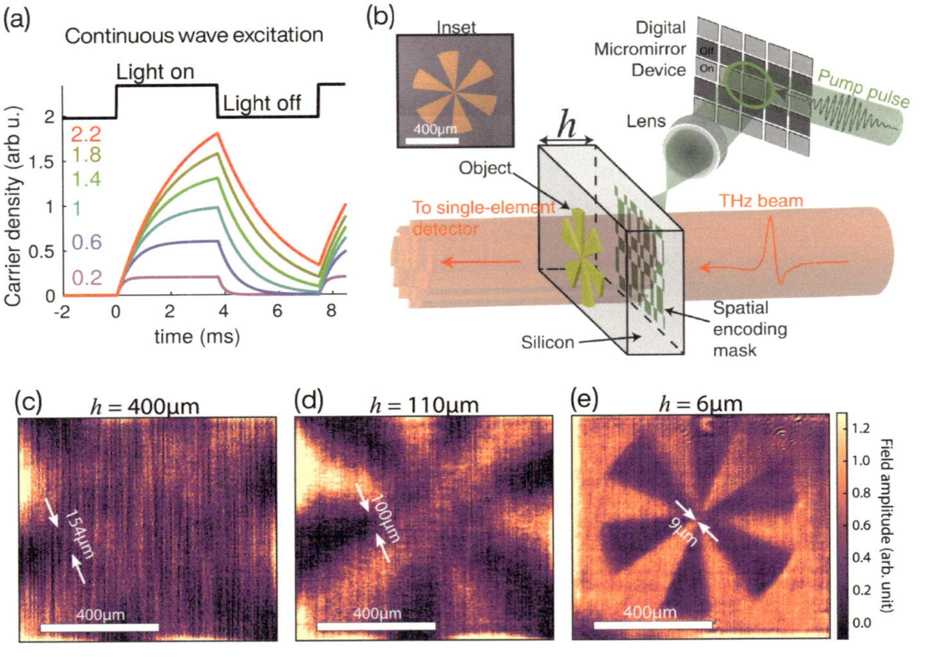

Figure 3.
(a) Carrier density (in arbitrary units) for different carrier lifetimes, shown by the colored numbers in ms, as we switch a continuous wave source on and off. (b) Illustration of imaging setup: Using a digital micromirror device and a lens, a pump pulse is spatially structured and projected onto a silicon wafer. This spatially modulates a coincident THz pulse. This THz pulse then passes through an object and is measured on a single-element THz detector. Inset is an optical image of a resolution test target (cartwheel) manufactured from gold on a 6 μm thick silicon wafer. (c-e) THz images of the cartwheel shown in inset in part (b) taken through 400, 110, 6μm thick silicon wafers respectively. See supplementary fig. S4 for a close-up annotated version of part c. note, the cartwheels in a, b have diameters larger than the field of view. THz polarization is horizontal in experiment figures (b-e) adapted from reference [53]. The Optical Society (OSA) publishing group copyright with re-use permission granted by Rayko I. Stantchev.

we plot the carrier density from Eq. (4) for different carrier lifetimes. It can be seen that for the larger τ_l values the carrier density still has not reached its ground state value of 0, however they reach larger carrier densities hence this will result in larger THz modulation for the same photo-excitation powers. Hooper et al. experimentally studied the excitation powers needed for efficient THz modulation with regards to carrier lifetimes in reference [54]. Further, the physical resolution with which a pattern can imprinted on the THz beam is going to be either limited by the resolution with which we project our visible-light pattern or the carrier diffusion dynamics. In most cases, photomodulators with lifetimes around tens of microseconds are used hence the diffusion lengths are going be hundreds of microns, which is much higher than the limits imposed by the projection of visible light. There is a thorough theoretical study of carrier diffusion and modulation depth for THz spatial light modulators in reference [55].

For pulsed optical-excitation probed by a synchronous THz pulse, Eq. (5) changes because the THz pulse can travel through the spatially photopumped region a few picoseconds after photoexcitation and for $\tau_l > 100ps$ then carrier recombination can be ignored[2] however I_0 then becomes power per pulse per unit area. Further, since the THz pulse can be set to arrive a few picoseconds after

[2] This is also assumes that the pulse repetition period is much longer than the carrier lifetime, ie. the carriers have relaxed back to their ground state by the time the next pulse arrives.

photoexictation, then the carrier diffusion lengths become very small (about 500 nm for silicon [53, 56]). This means that the fundamental resolution is set by the photoexicitation beam provided that one can access the THz near-fields before the THz beam has diffracted due to propagating through space. In practice, this means using ultra-thin photomodulators. **Figure 3(b)** show the setup of a study which shows how the THz image resolution is affected by the photomodulator thickness, with parts (c-e) showing the achieved resolution of 154, 100, 9 μm with silicon wafer thicknesses of 400, 110 and 6 μm achieving about $\lambda/45$ resolution. Recently Chen, et al. used an even thinner photomodulator, 180 nm thick vanadium dioxide, to obtain even better resolution at 4.5 mum at $\lambda/100$ [57].

3.2.2 Electrical modulators

Electrical based modulators are likely to be the long-term future solution for spatial THz modulators because they have very little fundamental limitations. Namely, the maximum switch-rates are limited by the carrier recombination rates meaning they can potentially achieve megahertz switch rates, provided the RC constants of the devices are taken into account, especially with electrically tunable materials such as graphene [58, 59]. Their size is determined by photolithographic manufacturing technologies, which is already orders of magnitudes smaller than the THz wavelengths meaning that pixel sizes can be highly subwavelength. In fact, sometimes THz modulation structures can be too large for some commercial photolithographic systems. They are fully self-contained and compact, which is their main advantage over the optical based modulators (see §3.2.1) that need a laser, a spatial visible-light modulator and a photomodulator. However, in the near-term future they are unlikely to be commercially available. This is because their modulation amplitudes are determined by how carriers are injected/depleted via electrical gating and this is yet to optimized.

One of the first demonstrations of this modulation technique was by Kleine-Ostmann et al. in 2004 [60] where they electronically depleted carriers from a GaAs/AlGaAs interface, achieving about 3% modulation across a broadband frequency range of 0.1 to 2 THz. Since then there have been numerous attempts at improving the modulation depth, see references [61, 62] for recent reviews. These efforts have included enhancing the interaction between the THz wave and the charge carrier regions by metamaterial structures [63, 64]. Others have recently used graphene as the modulator [65, 66]. Using metamaterials or Fabry-Perot type resonances to enhance the modulation depth has the trade-off of reducing the working frequencies of the modulator. A further note is that subwavelength grating structures can enhance the THz modulation over a broadband range [58] for the correct THz polarization.

In 2014 C. M. Watts et al. created an electrical based THz-SLM in reference [67], and **Figure 4(a)** shows their experimental schematic and part (b) shows an image of their SLM. They electrically change the THz absorption of a 2 μm thick layer of n-doped GaAs, with each individual pixel being addressed by a FPGA array. The authors study the noise in final images from raster, random and Hadamard masks. Some of their results can be seen in **Figure 4(c)** and **(d)** with part (e) showing their object. Note, they were able to obtain 8×8 images at 1frames-per-second with **Figure 4(f)** showing 5 frames of the object in part (e) moving across the field of view. The device itself does not have large modulation depths meaning it mostly serves as a powerful proof-of-concept device. Another study creating an electrical based THz-SLM used graphene to create a 4×4 electro spatial-modulator [68]. And at last but certainly not least, we have the 2020 contribution of Y. Malevich et al. [69] where they built a 16×16 SLM by electrically gating two graphene layers

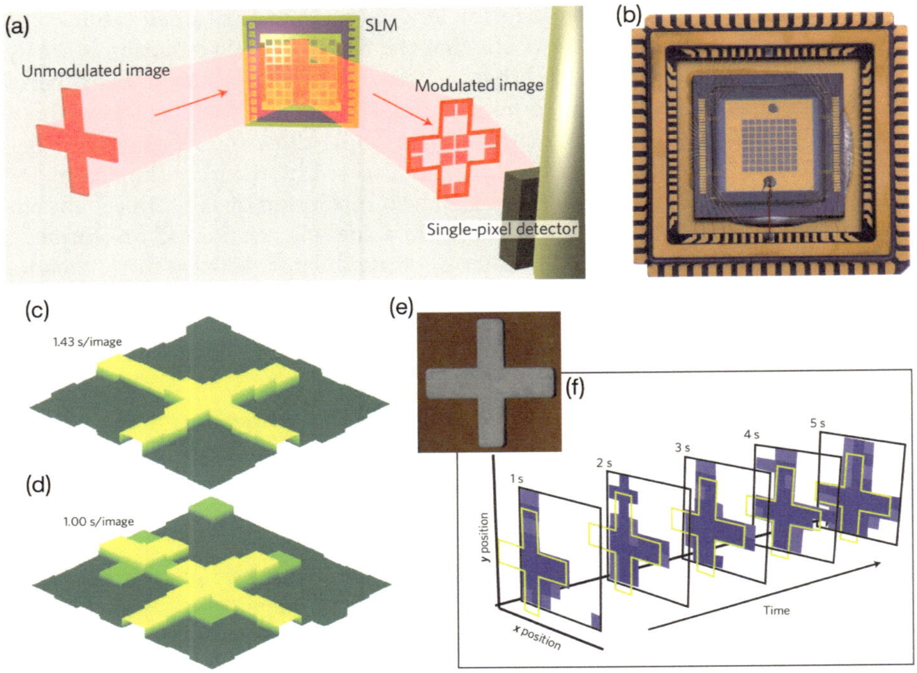

Figure 4.
(a) Schematic of the single-pixel imaging process utilizing an SLM. An image is spatially modulated by the metamaterial and the resulting radiation is sent to the single-pixel detector. (b) Photograph of the SLM (courtesy of K. burke, Boston College media technology services); total active area of the SLM is (4.8 mm²). (c) Image reconstruction using 64 masks with each mask displayed for 22.4 ms, giving a total image acquisition time of 1.43 s. (d) Image reconstruction using FISTA and 45 masks with each mask displayed for 22.4 ms, giving a total image acquisition time of 1 s. the colourmap scale of each reconstruction was chosen to best display the data. (e) Photograph of the object studied. The object was scanned across the field of view at a speed of 1.8 mm/s. (f) Consecutive tiles show FISTA reconstruction using 45 Hadamard masks. Only the first five frames are shown in the figure (see supplementary section 'compressive sensing' for full movie: 10 frames). The approximate position of the cross aperture is shown in yellow as a guide to the eye. Figure adapted from reference [67]. Copyright nature publishing group with re-use permission granted by Willie J. Padilla.

sandwiching an electrolyte. This device has broadband modulation depth and the electrolyte results in needing small voltages (below 4 V) meaning it can be easily integrated with microcontrollers. However the switch-rate, shown to be up-to 1 kHz, is limited by the generation of an electrical layer at the electrolyte-graphene interface, therefore more work is needed to optimize the modulation depth and switch rate. Nevertheless, these results [69] mark big progress towards a THz-SLM with a large number of pixels and low driving voltages.

3.3 Spatially patterned THz generation

Another innovative approach to single-pixel imaging is to create a spatially patterned beam at the generation step, rather than generate a homogeneous beam that is then spatially modulated, which has the benefit of not needing a THz-SLM. For the terahertz regime, this can be accomplished by three possible ways. First, having an array of photoconductive antennas [70–72], however this approach suffers from antenna cross-talk and inefficiency problems arising from the small working-area of the antennas whilst occupying a large area. Further, such antennas arrays have only been used as detectors. The second method is to use an electro-optic (EO) crystal that converts visible-light to THz frequencies via non-linear polarization effects [73]. The generation of THz radiation is localized to where the visible light is, hence projecting a spatially varying light beam will generate a THz-beam with the same spatial features. This idea was implemented by references [74, 75] where they used a used a SLM to pattern an 800 nm femtosecond pulse and project that onto a ZnTe crystal. The third method is similar to the second one, with the difference being the use of a spintronic THz emitter instead of an electro-optic crystal. Here the inverse spin Hall effect is used to generate an ultrafast current transient that generates the THz radiation [76–78]. The spatial patterning is again done by a visible-light SLM since the THz generation is again localized to areas where the optical-pump was shined upon. This was demonstrated by Chen et al. in reference [79].

The similarities between the spintronic and electro-optic crystal approaches are that they both require femtosecond pulses with mJ/cm^2 powers and both measure the THz field-strength with sub-picosecond temporal resolution. The differences come from limitations imposed by the electro-optic crystals. First, absorption at the EO crystal phonons limits the working frequencies whereas spintronic emitters can emit from 1 to 30 THz [77], although organic EO crystals can significantly improve working bandwidth with commercial companies offering crystals emitting from 0.3 to 16 THz [80]. Second, EO crystals are typically thicker than the THz wavelengths for good working efficiency whereas spintronic emitters are nanometer thick allowing easier access to the THz near-fields for subwavelength resolution [79]. Finally, for the spintronic emitters the emitted THz polarization state is easily controlled by the externally applied magnetic field needed for THz emission [76], whereas for the EO crystal approach one needs to add an additional visible light SLM [81] resulting in increased system complexity.

Figure 5(a) shows the experimental setup of reference [79] which uses the spintronic emitter array approach. Note that the EO crystal approach is identical in that you only have to replace the spintronic emitter with the EO crystal and remove the magnetic field, then place the object as close as possible to the emitter array. One should note that the use of the second DMD in **Figure 5(a)** is only to correct the phase front induced by first DMD, which can also be achieved by the technique shown in the supplementary information of reference [82]. The spintronic emitter is nanometer thick hence Chen et al. was able to resolve metallic lines 6 μm in width, as shown in **Figure 5(c)–(g)**, which is significantly shorter than the THz wavelengths. Note, at these subwavelength scales polarization effects become significant as be seen when comparing **Figure 5(d)** and **(e)** where they show images with horizontal and vertical polarization, respectively, of the same object.

3.4 Liquid crystal THz modulators

Liquid crystal modulators work by the re-orientating the material molecules under an applied voltage. As the molecules are oblong, this changes the refractive index that an electro-magnetic wave experiences. Therefore these devices are great for phase modulation giving the greatest freedom in the values that the sampling matrix can take. In other words, they can theoretically project a Fourier matrix that has grayscale complex-values. Note that complex valued masks can be used in conjunction with an intensity only detector to obtain an image that has phase and amplitude information [83]. A liquid crystal based SLM for THz was computationally studied [84]. However, the re-orientation of the molecules is a slow process, and in the visible light regime liquid crystal displays are typically limited to below 100 Hz switch rates. Due to the longer THz wavelengths, thicker layers of liquid crystals are needed resulting in even slower switch rates. For this reason, liquid crystal spatial modulators for THz radiation have been limited mostly to applications where slow switching speeds are acceptable such as dynamically controllable lenses [85, 86], absorption [87] or polarization control [88].

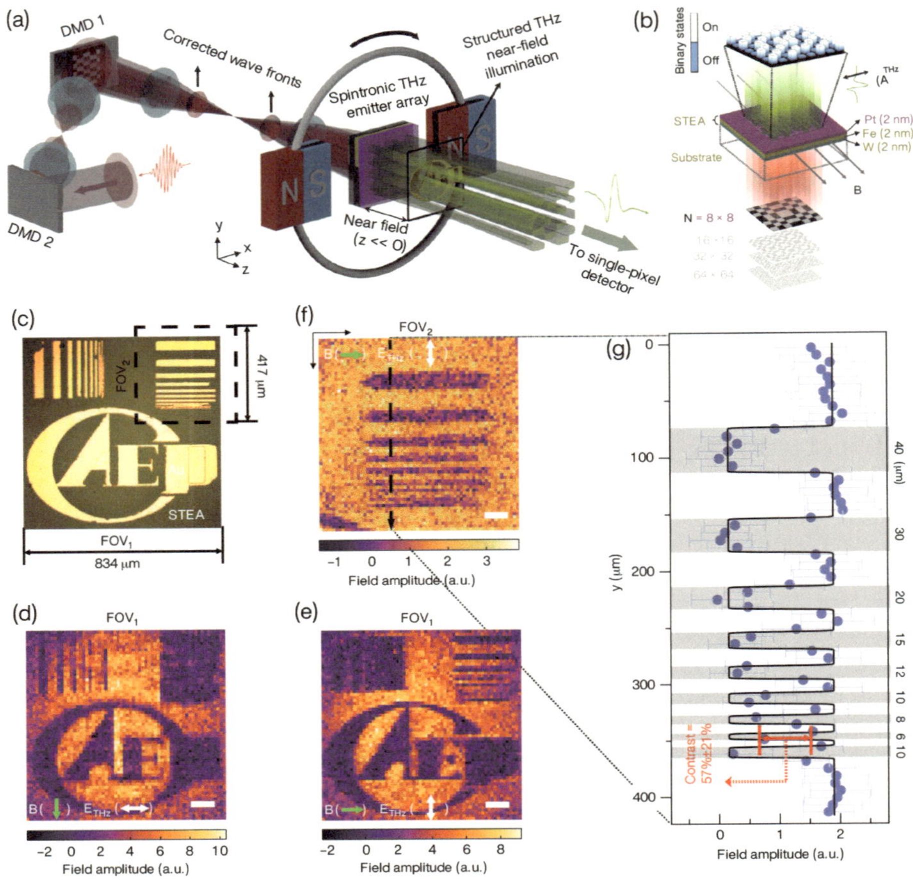

Figure 5.
(a) Schematic of the GHOSTEAM system. The spintronic THz emitter array (STEA) is excited by two-DMD-encoded fs laser pulses and generates spatially coded THz pulses. An object "CAEP" was placed in the near-field region ($z < < \lambda$). The illuminating THz pulse was collected and sent to a single-pixel detector. (b) Schematic of the STEA, consisting of a W(2 nm)/Fe(2 nm)/Pt(2 nm) trilayer heterostructure and working in the binary emission state with polarization perpendicular to applied magnetic field. (c) Optical image of an object with a field of view of FOV 1 = 834 μm 834 *μm.* the bright regions are gold attached on the 150-nm-thick protective SiO_2 *layer on top of the STEA. THz ghost images in FOV 1 with a magnetic field (green arrows) applied along the vertical direction (d) and horizontal direction (e). The pixel size and scale bar are 13.0* μm and 100 *μm, respectively, for both images. (f) THz ghost image in FOV 2 (indicated by the black dashed box in c) with a pixel size of 6.5 μm and a scale bar of 50 μm. the applied magnetic field B is along the horizontal direction (indicated by the green arrow), and the polarization of the THz radiation (indicated by the white double-headed arrow) is perpendicular to B. (g) Averaged amplitude of the THz field along the black dashed arrow in (f). Figure adapted and relabelled from reference [79]. Copyright nature publishing group with re-use permission under creative commons attribution 4.0 international license.*

4. Applications and discussion

The first demonstration of a single-pixel THz camera that uses a multi-pixel modulation approach[3] was in 2008 by Chan et al. [89]. Therein the authors showed amplitude and phase imaging was possible. Since Chan showed single-pixel THz imaging with metallic masks [89], most publications up until now have focused on

[3] Technically, a raster scanner is a single-pixel camera. However this chapter concerns itself when the beam of radiation has multiple scatters and apertures in each measurement. Therefore raster scanners are excluded.

improving the implementation by showing proof-of-concept modulation/generation techniques as opposed to potential applications. Shortly after in 2009 spectroscopic imaging was demonstrated [90]. The next experiment was in 2012 by Shen et al. [46] where they used a spinning disc with random masks, but it should be noted that their experiment used an infrared and a THz source whilst using the same SLM. The first demonstration of an optical based SLM was by Shrekenhamer et al. in 2013 [49]. The same group then published an electrical based SLM for single-pixel THz imaging in 2014 [67]. The next developments showed in 2016 that such imaging systems can detect a sub-wavelength fissure (8 μm) in a circuit board hidden by silicon [56]. Other applications show that these single-pixel cameras can be used for near-field biological imaging [91, 92], sub-wavelength THz mapping of graphene photoconductivity [82] as well as 2D tomographic imaging [93]. Further, observing hydration changes in a leaf due to intense light exposure over a 15 minute period has also been demonstrated [52]. Other recent studies include demonstrating a spinning disc system working from 3 to 13 THz using random masks [45]. Reference [94] shows how random, Hadamard and masks based on the discrete cosine transform perform in the THz regime and reference [95] studied Hadamard vs. Fourier masks in the context of undersampling performance.

The potential applications and capabilities of single-pixel THz cameras are directly determined by the THz source and detector, rather than the technique used to impart a spatial pattern in a beam of THz radiation. As such, it is unlikely for there to be a single-solution for all practical applications. Therefore, it is valuable to discuss where each of the techniques in §3 are likely to be used and how they compare to other THz imaging techniques. Metamaterial based SLMs, or specifically those that work over a few select frequencies, are unlikely to become widespread. This is because if you want to take an image only at one THz frequency then you can just use a bolometer detector array [18] and combine it with digital holography [19, 20] for an amplitude and phase measurement. Single-pixel imaging technologies are hampered by the serial data acquisition process, as such they will never truly compete at fast imaging rates with the parallel-data acquisition of detector arrays. As mentioned earlier, they compete in terms of being applicable when a detector array with the desired capabilities is just infeasible. This for THz technology translates itself to time-gated detection techniques that measure the amplitude and phase of a THz pulse with sub-picosecond temporal resolution, for example photo-conductive antennas [73].

The metallic/physical based masks, employed in §3.1, have contrast ratios of $\sim$ 99% over very large working frequencies. For example, holes in an aluminum plate could be used as an SLM for visible light as well for sub-THz frequencies and everything in between including ultrasound beams. The upper frequency limit is set by the plasma frequency whereas the lower frequency limit is set by the penetration depth at GHz frequencies. The major drawback is the physical movement involved in changing the masks. Using dynamically controlled mirrors, such those in **Figure 2**, requires no physical movement however their diffraction efficiency at low frequencies needs to be considered before implementation. Otherwise, their frequency response is going to be again to be determined by the reflection of a metal hence it is very broad. Such SLMs are probably most compatible for THz spectrometers that can measure between 1 and 30 THz [45], such as those based on air-plasma [96], spintronic [77] and organic EO crystals [80] emission techniques. This is of course compared to conductivity based SLMs of §3.2.

Modifying the Drude plasma frequency of a semiconductor via injection/depletion of charge carriers has great potential for compact integration of the entire imaging system, as long as electrical gating is used as it negates the need for an extra pump-laser. The drawback is that to ensure modulation depth over a large frequency range the Drude plasma frequency has to be sufficiently modified. For example, after photoexcitation of silicon if ω_p moves from 0.01 THz to 25 THz, then the modulation depth for frequencies above 20 THz is going to be very poor. However, if the measurement only needs to measure frequencies between 0.1 and 2 THz with 100 fs temporal resolution, the working frequencies of many commercial THz-TDS systems, then this plasma frequency is going to be fine. For example, reference [52] achieved a 32 × 32 THz video with 6 frames-per-second using a system with such measurement capabilities. Their SLM is an optical excitation one, hence their system is not as compact as that shown in **Figure 4** but it demonstrates that fast THz imaging with single-pixel cameras is possible. Nevertheless, Chen et al. showed a THz amplitude modulation of near 200%[4] over the frequency range of 0.1 to 2 THz using electrical gating of graphene [59]. This shows electrical modulators have great potential but they are far from being commercially available as it was only in 2020 that a demonstration of 16 × 16 THz modulator was published [69], which has lower modulation depth and slower switch-rate than reference [52]. Finally, the physical resolution of these approaches is ultimately limited by the visible light pattern or the manufacturing technology used to make the array. As such, electrical modulators are capable of reaching resolutions of 150 nm and then not have issues of misalignment. In terms of sheer resolution, however, they are unlikely to compete with the 10 nm resolution achieved with AFM-tips [24] although due to the multiple scatterers and apertures in each measurement they will have larger SNRs [56] and thus offer quicker acquisition. A side note is that using Hadamard-Sylvester masks at sub-wavelength resolutions can result in grating-like diffraction effects for individual masks as shown in chapter 5.3.2 of reference [33].

Direct generation of a spatially varying THz beam, §3.3, has the benefit of not needing a THz-SLM at all, reducing the system complexity, whilst having visible light resolution. However, all the current techniques demonstrated have one major drawback: they need pulse energies around 1 mJ/cm^2 for sufficient signal-to-noise. A back of the envelope calculation shows that if such fluences are needed over a 500 × 500 μm square, then average laser powers of 2.5 mW, 0.625 W and 125 W incident onto the emitter area would be needed if the laser repetition rates are 1 kHz, 250 kHz and 50 MHz respectively[5]. The lower repetition rate lasers have lower signal-to-noise (measuring less pulses per unit time) and usually have larger laser noise-fluctuations causing further degradation. As such, these techniques are likely only usable for areas smaller than 500 × 500 μm where accessing the THz near-fields is necessary. Their competing technology is mapping out the THz fields onto a EO crystal using a CCD camera [21], which can achieve near-field resolution and still measure the THz-fields with sub-picosecond temporal resolution. Theoretically, neither the technique of reference [21] and that shown in **Figure 5** have an edge over of the other. However, in terms of experimental implementation the spintronic emitter array [79] has some advantages. Namely, easily controllable polarization and you need a high-quality camera for the EO-mapping approach [21] wheres as DMDs are low cost and found in everyday projectors. Finally, the EO-mapping approach needs an ultra-thin EO crystal thereby necessitating the use of an intense THz-source, whereas the spintronic array could possibly be implemented with a 100 kHz laser which should improve the SNR.

---

[4] This is was done by changing the reflection coefficient from 1 to −1 near the Brewster angle.

[5] Losses from various optical elements would occur and hence higher laser powers will be needed initially.

5. Conclusions

Although the first experimental implementations of single-pixel cameras can be traced back to 1976 [97], such imaging approaches were not widely studied or implemented in the commercial world. The reason is that the serial measurement of such ideas can not compete with parallel data acquisition of imaging arrays. Further, compressed sensing techniques began gaining mainstream attention in 2006 after two publications [37, 38]. This coincides with the development of visible light spatial modulators thereby allowing the implementation of the ideas in references [37, 38]. Whilst inherently slower than imaging arrays, these single-pixel cameras are much more robust and easier to implement in areas where imaging array technology is unavailable. In particular, the terahertz frequency regime.

This book chapter began by outlining the current state of THz cameras. Then it discusses the background theory of single-pixel imaging techniques. Most of the chapter was dedicated to discussing the current state of spatial THz-light modulators for use in single-pixel THz imaging in §3. The modulation techniques discussed are based on mechanical masks §3.1, optical §3.2.1 and electrical §3.2.2 modulation of conductivity, direct generation of spatially varying THz beams §3.3 and liquid crystal modulators §3.4. The final §4 is a discussion of the current state of single-pixel cameras as well the pros and cons of each different THz-SLM methodology.

The most advanced THz-SLM at present are those based on optical excitation of semiconductors, §3.2.1, however this is mostly because of their ease of implementation with current technology. Electrical modulators, §3.2.2, are likely to become the most used in the long term because compared to optical modulators they offer the same potential switch-speeds if not quicker, the same working frequencies, they are more compact and the physical resolution is limited by photolithrographic technology as opposed to visible light resolution. Further, the physical resolution been limited by the diffraction of the visible light-pump in an optical SLM is only achievable with high-power low-rep rate lasers, whereas electrical modulators would be compatible with fiber-based THz systems, that have magnitudes higher SNRs and are much cheaper, whilst maintaining the same resolution. Generating a THz beam with spatial variations has great resolution set by the visible-light diffraction limit, however their appeal will significantly drop the moment someone builds an electrical-based THz modulator with say 500×500 nm sized array elements that can be placed in the near-field of an object. Mechanical masks are mostly likely to be used in some niche applications that for example require simultaneous imaging with a THz beam and an ultrasonic beam. Liquid crystal modulators are unlikely to be used in single-pixel imaging unless there is a breakthrough that allows their switch-rates to reach 10 kHz, although they offer the most freedom in regards to the values of the projected masks (ie. complex grayscale values).

Ultimately, the development of single-pixel THz cameras is likely to proceed with optical modulators being used in university laboratories to optimize the algorithms and methodologies used in image recovery as well as synchronization of all the equipment. Simultaneously there will be an effort to develop electrical based array modulators that have fast-switch rates and large modulation depth over a broadband frequency range. Then the miniaturization of such modulator arrays will start and it is likely that at this point such commercially available THz-SLMs will become available from new specialized start-up companies. Spatially-generated THz beams will likely remain only in laboratories for fundamental studies of different systems, but are unlikely to be used for industrial and commercial applications mostly due to the requirement of pump powers of ~ 1 mJ/cm^2. Further, they are compatible with time-gated synchronization of visible-pump THz-probe experiments allowing the study of sub-picosecond photoexcitation dynamics, hence their appeal to university labs.

Acknowledgements

This work was partially supported by the Research Grants Council of Hong Kong (project numbers 14206717 and 14201415), The Hong Kong Innovation and Technology Fund (project number ITS/371/16), The Engineering and Physical Sciences Research Council (grant number EP/S021442/1), and the Royal Society Wolfson Merit Award (EPM).

Conflict of interest

The authors declare no conflict of interest.

Nomenclature

AFM	Atomic force microscope
PCB	Printed Circuit Board
FPGA	Field-programmable gate array
SLM	Spatial light modulator
THz	Terahertz
EO	Electro-optic
SNR	Signa-to-noise ratio
DMD	Digital micromirror device
VO_2	Vanadium Dioxide
CCD	Charge Coupled device

Author details

Rayko Ivanov Stantchev[1] and Emma Pickwell-MacPherson[1,2]*

1 Department of Electronic Engineering, The Chinese University of Hong Kong, Hong Kong SAR

2 Physics Department, Warwick University, Coventry, UK

*Address all correspondence to: rayko@cuhk.edu.hk

References

[1] Mittleman, D. M. Twenty years of terahertz imaging [Invited]. *Optics Express* DOI: 10.1364/oe.26.009417 (2018).

[2] Chan, W. L., Deibel, J. & Mittleman, D. M. Imaging with terahertz radiation. *Reports on Progress in Physics* **70**, 1325–1379, DOI: 10.1088/0034-4885/70/8/R02 (2007).

[3] Jansen, C. *et al.* Terahertz imaging: applications and perspectives. *Applied Optics* 49, E48, DOI: 10.1364/AO.49.000E48 (2010).

[4] Dhillon, S. S. *et al.* The 2017 terahertz science and technology roadmap, DOI: 10.1088/1361-6463/50/4/043001 (2017).

[5] Dong, J. *et al.* Terahertz frequency-wavelet domain deconvolution for stratigraphic and subsurface investigation of art painting. *Optics Express* **24**, 26972–26985, DOI: 10.1364/OE.24.026972 (2016).

[6] Ahi, K., Asadizanjani, N., Shahbazmohamadi, S., Tehranipoor, M. & Anwar, M. Terahertz characterization of electronic components and comparison of terahertz imaging with x-ray imaging techniques. *SPIE* 94830K, DOI: 10.1117/12.2183128 (2015).

[7] Karpowicz, N. *et al.* Compact continuous-wave subterahertz system for inspection applications. *Applied Physics Letters* **86**, 054105, DOI: 10.1063/1.1856701 (2005).

[8] Dong, J. *et al.* Visualization of subsurface damage in woven carbon fiber-reinforced composites using polarization-sensitive terahertz imaging. *NDT and E International* DOI: 10.1016/j.ndteint.2018.07.001 (2018).

[9] Ulbricht, R., Hendry, E., Shan, J., Heinz, T. F. & Bonn, M. Carrier dynamics in semiconductors studied with time-resolved terahertz spectroscopy. *Reviews of Modern Physics* **83**, 543–586, DOI: 10.1103/RevModPhys.83.543 (2011).

[10] Shen, Y. C. Terahertz pulsed spectroscopy and imaging for pharmaceutical applications: A review. *International Journal of Pharmaceutics* **417**, 48–60, DOI: 10.1016/j.ijpharm.2011.01.012 (2011).

[11] Konek, C., Wilkinson, J., Esenturk, O., Heilweil, E. & Kemp, M. Terahertz spectroscopy of explosives and simulants: RDX, PETN, sugar, and L-tartaric acid. 73110K, DOI: 10.1117/12.817913 (International Society for Optics and Photonics, 2009).

[12] Kampfrath, T., Tanaka, K. & Nelson, K. A. Resonant and nonresonant control over matter and light by intense terahertz transients. *Nature Photonics* 7, 680–690, DOI: 10.1038/nphoton.2013.184 (2013).

[13] Yu, C., Fan, S., Sun, Y. & Pickwell-MacPherson, E. The potential of terahertz imaging for cancer diagnosis: A review of investigations to date. *Quantitative Imaging in Medicine and Surgery* **2**, 33–45, DOI: 10.3978/j.issn.2223-4292.2012.01.04 (2012).

[14] Hernandez-Cardoso, G. G. *et al.* Terahertz imaging for early screening of diabetic foot syndrome: A proof of concept. *Scientific Reports* 7, 1–9, DOI: 10.1038/srep42124 (2017).

[15] Titova, L. V. *et al.* Intense THz pulses cause H2AX phosphorylation and activate DNA damage response in human skin tissue. *Biomedical Optics Express* DOI: 10.1364/boe.4.000559 (2013).

[16] Escorcia, I., Grant, J., Gough, J. & Cumming, D. R. S. Uncooled CMOS terahertz imager using a metamaterial

absorber and pn diode. *Optics Letters* **41**, 3261–3264, DOI: 10.1364/ol.41.003261 (2016).

[17] Rogalski, A. & Sizov, F. Terahertz detectors and focal plane arrays. *Opto-Electronics Review* **19**, 346–404, DOI: 10.2478/s11772-011-0033-3 (2011).

[18] Amanti, M. I., Scalari, G., Beck, M. & Faist, J. Stand-alone system for high-resolution, real-time terahertz imaging. *Optics Express* **20**, 2772–2778, DOI: 10.1364/oe.20.002772 (2012).

[19] Locatelli, M. *et al.* Real-time terahertz digital holography with a quantum cascade laser. *Scientific Reports* **5**, 13566, DOI: 10.1038/srep13566 (2015).

[20] Yamagiwa, M. *et al.* Real-Time Amplitude and Phase Imaging of Optically Opaque Objects by Combining Full-Field Off-Axis Terahertz Digital Holography with Angular Spectrum Reconstruction. *Journal of Infrared, Millimeter, and Terahertz Waves* **39**, 561–572, DOI: 10.1007/s10762-018-0482-6 (2018).

[21] Blanchard, F. *et al.* Real-time terahertz near-field microscope. *Optics Express* **19**, 8277, DOI: 10.1364/OE.19.008277 (2011).

[22] Usami, M. *et al.* Development of a THz spectroscopic imaging system. *Physics in Medicine and Biology* **47**, 3749–3753, DOI: 10.1088/0031-9155/47/21/311 (2002).

[23] Huber, A. J., Keilmann, F., Wittborn, J., Aizpurua, J. & Hillenbrand, R. Terahertz near-field nanoscopy of mobile carriers in single semiconductor nanodevices. *Nano Letters* DOI: 10.1021/nl802086x (2008).

[24] Eisele, M. *et al.* Ultrafast multi-terahertz nano-spectroscopy with sub-cycle temporal resolution. *Nature Photonics* DOI: 10.1038/nphoton.2014.225 (2014).

[25] van der Valk, N. C. J. & Planken, P. C. M. Electro-optic detection of subwavelength terahertz spot sizes in the near field of a metal tip. *Applied Physics Letters* **81**, 1558, DOI: 10.1063/1.1503404 (2002).

[26] Macfaden, A. J., Reno, J. L., Brener, I. & Mitrofanov, O. 3 μm aperture probes for near-field terahertz transmission microscopy. *Applied Physics Letters* **104**, 011110, DOI: 10.1063/1.4861621 (2014).

[27] Chiu, C.-M. *et al.* All-terahertz fiber-scanning near-field microscopy. *Optics Letters* **34**, 1084, DOI: 10.1364/OL.34.001084 (2009).

[28] Hunsche, S., Koch, M., Brener, I. & Nuss, M. THz near-field imaging. *Optics Communications* **150**, 22–26, DOI: 10.1016/S0030-4018(98)00044-3 (1998).

[29] Zhao, J. *et al.* Terahertz imaging with sub-wavelength resolution by femtosecond laser filament in air. *Scientific Reports* DOI: 10.1038/srep03880 (2014).

[30] Gibson, G. M., Johnson, S. D. & Padgett, M. J. Single-pixel imaging 12 years on: a review. *Optics Express* **28**, 28190, DOI: 10.1364/OE.403195 (2020).

[31] Kutyniok, G. Theory and applications of compressed sensing. *GAMM-Mitteilungen* **36**, 79–101, DOI: 10.1002/gamm.201310005 (2013).

[32] Yates, F. Complex Experiments. *Supplement to the Journal of the Royal Statistical Society* **8**, 27–41, DOI: 10.2307/2983638 (1935).

[33] Stantchev, R. I. *Non-invasive near-field THz imaging using a single pixel detector*. Ph.D. thesis, University of Exeter (2017).

[34] Zhang, Z., Wang, X., Zheng, G. & Zhong, J. Hadamard single-pixel

imaging versus Fourier single-pixel imaging. *Optics Express* DOI: 10.1364/oe.25.019619 (2017).

[35] Harwit, M. & Sloane, N. J. A. *Hadamard Transform Optics* (Academic Press, New York, 1979).

[36] Paley, R. E. A. C. On orthogonal matrices. *J. Math. Phys* **12**, 311–320 (1933).

[37] Candes, E., Romberg, J. & Tao, T. Robust uncertainty principles: exact signal reconstruction from highly incomplete frequency information. *IEEE Transactions on Information Theory* **52**, 489–509, DOI: 10.1109/TIT.2005.862083 (2006).

[38] Donoho, D. L. Compressed sensing. *IEEE Transactions on Information Theory* **52**, 1289–1306, DOI: Doi10.1109/Tit.2006.871582 (2006).

[39] Candes, E. & Romberg, J. l1-magic: Recovery of sparse signals via convex programming, https://github.com/scgt/l1magic (2005).

[40] Li, C., Yin, W., Jiang, H. & Zhang, Y. An efficient augmented Lagrangian method with applications to total variation minimization. *Computational Optimization and Applications* DOI: 10.1007/s10589-013-9576-1 (2013).

[41] Li, C., Yin, W. & Yin, Z. TVAL3: TV minimization by Augmented Lagrangian and ALternating direction ALgorithms, https://www.caam.rice.edu/~optimization/L1/TVAL3/ (2010).

[42] Czajkowski, K. M., Pastuszczak, A. & Kotyński, R. Real-time single-pixel video imaging with Fourier domain regularization. *Optics Express* **26**, 20009, DOI: 10.1364/OE.26.020009 (2018).

[43] Czajkowski, K. M., Pastuszczak, A. & Kotyński, R. FDRI single-pixel imaging, https://github.com/KMCzajkowski/ FDRI-single-pixel-imaging (2020).

[44] Giles, R. H. Characterization of Material Properties at Terahertz Frequencies. *Submillimeter Technology Laboratory, University of Massachusetts* (1995).

[45] Vallés, A., Jiahuan, H. E., Ohno, S., Omatsu, T. & Miyamoto, K. Broadband high-resolution terahertz single-pixel imaging. *arXiv* DOI: 10.1364/oe.404143 (2020).

[46] Shen, H. *et al.* Spinning disk for compressive imaging. *Optics Letters* DOI: 10.1364/ol.37.000046 (2012).

[47] Kappa, J. *et al.* Electrically Reconfigurable Micromirror Array for Direct Spatial Light Modulation of Terahertz Waves over a Bandwidth Wider Than 1 THz. *Scientific Reports* DOI: 10.1038/s41598-019-39152-y (2019).

[48] Alius, H. & Dodel, G. Amplitude-, phase-, and frequency modulation of far-infrared radiation by optical excitation of silicon. *Infrared Physics* 32, 1–11, DOI: 10.1016/0020-0891(91)90090-3 (1991).

[49] Shrekenhamer, D., Watts, C. M. & Padilla, W. J. Terahertz single pixel imaging with an optically controlled dynamic spatial light modulator. *Optics Express* **21**, 12507, DOI: 10.1364/OE.21.012507 (2013).

[50] Cavalleri, A. *et al.* Femtosecond structural dynamics in VO2 during an ultrafast solid-solid phase transition. *Physical Review Letters* DOI: 10.1103/physrevlett.87.237401 (2001).

[51] Jeong, J. *et al.* Suppression of metal-insulator transition in VO2 by electric field-induced oxygen vacancy formation. *Science* DOI: 10.1126/science.1230512 (2013).

[52] Stantchev, R. I., Yu, X., Blu, T. & Pickwell-MacPherson, E. Real-time terahertz imaging with a single-pixel

detector. *Nature Communications* DOI: 10.1038/s41467-020-16370-x (2020).

[53] Stantchev, R. I. *et al.* Compressed sensing with near-field THz radiation. *Optica* DOI: 10.1364/optica.4.000989 (2017).

[54] Hooper, I. R. *et al.* High efficiency photomodulators for millimeter wave and THz radiation. *Scientific Reports* DOI: 10.1038/s41598-019-54011-6 (2019).

[55] Kannegulla, A., Shams, M. I. B., Liu, L. & Cheng, L.-J. Photo-induced spatial modulation of THz waves: opportunities and limitations. *Optics Express* DOI: 10.1364/oe.23.032098 (2015).

[56] Stantchev, R. I. *et al.* Noninvasive, near-field terahertz imaging of hidden objects using a single-pixel detector. *Science Advances* 2, e1600190 (2016).

[57] Chen, S.-C. *et al.* Terahertz wave near-field compressive imaging with a spatial resolution of over $\lambda/100$. *Optics Letters* DOI: 10.1364/ol.44.000021 (2019).

[58] Sun, Y. *et al.* Graphene-loaded metal wire grating for deep and broadband THz modulation in total internal reflection geometry. *Photonics Research* DOI: 10.1364/prj.6.001151 (2018).

[59] Chen, Z. *et al.* Graphene controlled Brewster angle device for ultra broadband terahertz modulation. *Nature Communications* DOI: 10.1038/s41467-018-07367-8 (2018).

[60] Kleine-Ostmann, T., Dawson, P., Pierz, K., Hein, G. & Koch, M. Room-temperature operation of an electrically driven terahertz modulator. *Applied Physics Letters* DOI: 10.1063/1.1723689 (2004).

[61] Degl'Innocenti, R., Kindness, S. J., Beere, H. E. & Ritchie, D. A. All-integrated terahertz modulators. *Nanophotonics* DOI: 10.1515/nanoph-2017-0040 (2018).

[62] Wang, L. *et al.* A review of THz modulators with dynamic tunable metasurfaces. *Nanomaterials* DOI: 10.3390/nano9070965 (2019).

[63] Chen, H. T. *et al.* Active terahertz metamaterial devices. *Nature* DOI: 10.1038/nature05343 (2006).

[64] Chen, H. T. *et al.* A metamaterial solid-state terahertz phase modulator. *Nature Photonics* DOI: 10.1038/nphoton.2009.3 (2009).

[65] Lee, S. H. *et al.* Switching terahertz waves with gate-controlled active graphene metamaterials. *Nature Materials* DOI: 10.1038/nmat3433 (2012).

[66] Sensale-Rodriguez, B. *et al.* Extraordinary control of terahertz beam reflectance in graphene electro-absorption modulators. *Nano Letters* DOI: 10.1021/nl3016329 (2012).

[67] Watts, C. M. *et al.* Terahertz compressive imaging with metamaterial spatial light modulators. *Nat. Photon.* **8**, 605–609, DOI: 10.1038/nphoton.2014.139 (2014).

[68] Sensale-Rodriguez, B. *et al.* Terahertz imaging employing graphene modulator arrays. *Optics Express* DOI: 10.1364/oe.21.002324 (2013).

[69] Malevich, Y., Ergoktas, S., M., Bakan, G., Steiner, P. & Kocabas, C. Video-Speed Graphene Modulator Arrays for Terahertz Imaging Applications. *ACS Photonics* 7, 2374–2380, DOI: 10.1021/acsphotonics.0c00991 (2020).

[70] Nallappan, K. *et al.* A dynamically reconfigurable terahertz array antenna for 2D-imaging applications. In *2017 Photonics North, PN 2017,* DOI: 10.1109/PN.2017.8090603 (2017).

[71] Pradarutti, B. *et al.* Terahertz line detection by a microlens array coupled photoconductive antenna array. *Optics Express* DOI: 10.1364/oe.16.018443 (2008).

[72] Brahm, A. *et al.* Multichannel terahertz time-domain spectroscopy system at 1030 nm excitation wavelength. *Optics Express* DOI: 10.1364/oe.22.012982 (2014).

[73] Reimann, K. Table-top sources of ultrashort THz pulses. *Reports on Progress in Physics* DOI: 10.1088/0034-4885/70/10/R02 (2007).

[74] Olivieri, L. *et al.* Hyperspectral terahertz microscopy via nonlinear ghost imaging. *Optica* 7, 186–191, DOI: 10.1364/OPTICA.381035 (2020).

[75] Zhao, J., Yiwen, E., Williams, K., Zhang, X. C. & Boyd, R. W. Spatial sampling of terahertz fields with sub-wavelength accuracy via probe-beam encoding. *Light: Science and Applications* DOI: 10.1038/s41377-019-0166-6 (2019).

[76] Kong, D. *et al.* Broadband Spintronic Terahertz Emitter with Magnetic-Field Manipulated Polarizations. *Advanced Optical Materials* DOI: 10.1002/adom.201900487 (2019).

[77] Seifert, T. *et al.* Efficient metallic spintronic emitters of ultrabroadband terahertz radiation. *Nature Photonics* DOI: 10.1038/nphoton.2016.91 (2016).

[78] Kampfrath, T. *et al.* Terahertz spin current pulses controlled by magnetic heterostructures. *Nature Nanotechnology* DOI: 10.1038/nnano.2013.43 (2013).

[79] Chen, S. C. *et al.* Ghost spintronic THz-emitter-array microscope. *Light: Science and Applications* DOI: 10.1038/s41377-020-0338-4 (2020).

[80] Photonics, R. Terahertz Generators and Detectors: Electro-Optic DSTMS Crystals, http://rainbowphotonics.com/prod_dstms.php.

[81] Sato, M. *et al.* Terahertz polarization pulse shaping with arbitrary field control. *Nature Photonics* DOI: 10.1038/nphoton.2013.213 (2013).

[82] Hornett, S. M., Stantchev, R. I., Vardaki, M. Z., Beckerleg, C. & Hendry, E. Subwavelength Terahertz Imaging of Graphene Photoconductivity. *Nano Letters* DOI: 10.1021/acs.nanolett.6b03168 (2016).

[83] Saqueb, S. A. N. & Sertel, K. Phase-Sensitive Single-Pixel THz Imaging Using Intensity-Only Measurements. *IEEE Transactions on Terahertz Science and Technology* DOI: 10.1109/TTHZ.2016.2610760 (2016).

[84] Savo, S., Shrekenhamer, D. & Padilla, W. J. Liquid crystal metamaterial absorber spatial light modulator for THz applications. *Advanced Optical Materials* DOI: 10.1002/adom.201300384 (2014).

[85] Zhou, S. *et al.* Liquid crystal integrated metalens with dynamic focusing property. *Optics Letters* DOI: 10.1364/ol.398601 (2020).

[86] Shen, Z. *et al.* Liquid crystal tunable terahertz lens with spin-selected focusing property. *Optics Express* DOI: 10.1364/oe.27.008800 (2019).

[87] Shrekenhamer, D., Chen, W. C. & Padilla, W. J. Liquid crystal tunable metamaterial absorber. *Physical Review Letters* DOI: 10.1103/PhysRevLett.110.177403 (2013).

[88] Chen, X. *et al.* Highly Efficient Ultra-Broadband Terahertz Modulation Using Bidirectional Switching of Liquid Crystals. *Advanced Optical Materials* DOI: 10.1002/adom.201901321 (2019).

[89] Chan, W. L. *et al.* A single-pixel terahertz imaging system based on

compressed sensing. *Applied Physics Letters* **93**, 121105, DOI: 10.1063/1.2989126 (2008).

[90] Shen, Y. C. *et al.* Terahertz pulsed spectroscopic imaging using optimized binary masks. *Applied Physics Letters* DOI: 10.1063/1.3271030 (2009).

[91] Stantchev, R. I. *et al.* Subwavelength hyperspectral THz studies of articular cartilage. *Scientific Reports* DOI: 10.1038/s41598-018-25057-9 (2018).

[92] Barr, L. E. *et al.* Total internal reflection based super-resolution imaging for sub-IR frequencies. *arXiv preprint: 2006.02091* (2020).

[93] Mohr, T., Herdt, A. & Elsässer, W. 2D tomographic terahertz imaging using a single pixel detector. *Optics Express* DOI: 10.1364/oe.26.003353 (2018).

[94] Augustin, S., Frohmann, S., Jung, P. & Hübers, H. W. Mask responses for single-pixel terahertz imaging. *Scientific Reports* DOI: 10.1038/s41598-018-23313-6 (2018).

[95] She, R., Liu, W., Lu, Y., Zhou, Z. & Li, G. Fourier single-pixel imaging in the terahertz regime. *Applied Physics Letters* DOI: 10.1063/1.5094728 (2019).

[96] Clough, B., Dai, J. & Zhang, X. C. Laser air photonics: Beyond the terahertz gap, DOI: 10.1016/S1369-7021(12)70020-2 (2012).

[97] Sloane, N. J. A. & Harwit, M. Masks for Hadamard transform optics, and weighing designs. *Applied Optics* DOI: 10.1364/ao.15.000107 (1976).

Chapter 4

Probing Biological Water Using Terahertz Absorption Spectroscopy

Rajib Kumar Mitra and Dipak Kumar Palit

Abstract

Hydrogen bonding properties of water molecules, which are confined in microcavities of biological interfaces, are significantly different from those of bulk water and drive most of the complex biological processes. While NMR, X-ray and UV–vis-IR spectroscopic techniques have been found inadequate for describing the dynamics of the thick (20–40 Å) sheath of hydration layer around biomolecules, recently developed THz spectroscopy has emerged as a powerful technique to directly probe the collective dynamics of hydrogen bonds in the hydration layer, which control all important functions of the biomolecules in life. Both laser and accelerator-based THz sources are intense enough to penetrate up to about 100 μm thick water samples, which makes THz transmission and/or dielectric relaxation measurements possible in aqueous solutions. These measurements provide valuable information about the rattling and rotational motions of hydrated ions, making, breaking and rearrangement of hydrogen bonds in hydration layer as well as hydrophilic and hydrophobic interactions between biomolecule and water. THz spectroscopy has also been successfully applied to study the effect of modulation of the physical conditions, like temperature, pH, concentration of proteins and chemical additives, on the structure and dynamics of hydration layer. THz spectroscopy has also been applied to study the processes of denaturation, unfolding and aggregation of biomolecules.

Keywords: Hydration Layer, Biomolecules, Hydrogen bond reorganization dynamics, THz spectroscopy, Protein aggregation

1. Introduction

Water molecules, which reside on the surfaces of proteins or lipid bilayers or in tissues and cells, exhibit properties that are significantly different from those found in pure or bulk water as water molecules in such biological systems face additional interactions [1–7]. Unique properties of such water molecules, which are confined in microenvironments of biological surfaces or interfaces are popularly termed as 'biological water'. This water drives many biological processes, in which it plays wide varieties of roles at different levels of complexity making its participation increasingly evident as an active agent and not simply as the spectator solvent [8–12]. Biological water differs from bulk water in a number of ways. First, clustering of the water molecules at the surface of a protein increases the local density by as much as 25% compared to that in bulk water [13, 14]. MD simulation by Smith

IntechOpen

et al. has revealed that about half of this density increase arises from the shortening of the average water – water distances and the other half from an increase in the coordination number [15]. Second, replacement of the water–water hydrogen bonds by the water–protein hydrogen bonds in the protein hydration layer lowers the freezing point of the hydration water (i.e. prevents formation of ice). The hydration layer of many proteins does not freeze even at sub-zero temperatures and thus life may sustain even at low temperatures [16–18]. Third, in bulk water, mutual polarization of the hydrogen bonded water molecules increases the dipole moment and dielectric constant. Such polarization is absent for water molecules in hydration layers of biological molecules, i.e. biological water is less polar than bulk water [19]. Further, measurements of water dynamics suggest that around 10–25% of water molecules in cells have slower reorientation dynamics, by around an order of magnitude, than those in the bulk [2, 20]. Centrality of liquid water to life can be easily understood from the fact that condition for the search for the possibility of life elsewhere is the presence of trace of water [21]. In spite of this status, role of water in sustaining life is still not understood perfectly [12]. It is now, however, clear that water plays an active role in the life of the cell over many scales of time and distance and exhibits diverse structural and dynamical roles in molecular cell biology [3, 22].

In liquid (or bulk) water, the molecules form a tetrahedrally coordinated motif, which is the building block of ephemeral six-membered (ice-like) or five-membered (clathrate-like) ring structure, consisting of fluctuating network of hydrogen bonds, but each bond has an average lifetime of about a picosecond [23]. Thermodynamics of hydration in water are generally governed by a balance between the enthalpic and entropic consequences, namely the enthalpies of water–water and water–solute interactions (hydrogen bonding, electrostatic, and van der Waals) and the entropies of disrupting the relatively ordered hydrogen-bonded networks of bulk water and forming new hydrogen bonds to suit the geometric factors of the biological interfaces [15, 24, 25].

Similarities in water dynamics in hydration shells of various proteins [26] suggest that the dynamics are determined by rather general features of surface chemistry and topology, which induce excluded volume effects and hinder the approach of new hydrogen bond acceptors within the hydration network. There is now universal recognition that the dynamical behavior of biological macromolecules cannot be decoupled from that of water and the dynamical behavior of the hydration shell largely controls the chemical function of the biological molecules [27, 28].

Since the development of this understanding that biomolecules are surrounded by a sheath of hydration water, which takes active part in all of their normal activities, extensive efforts have been made to perceive the detailed structure and dynamics of the hydration layer using a variety of spectroscopic methods, such as X-ray and neutron scattering [14, 29, 30], NMR [31–33], second harmonic generation [34, 35] and ultrafast fluorescence and IR spectroscopies [1, 2, 22, 26, 36, 37], assisted by ab initio and molecular dynamics simulations [24, 25, 38, 39]. However, these measurements could provide information about the dynamics of water and the biomolecules at a low-hydration level corresponding to the first layer of hydration water only. Since fluctuations of protein and solvent dynamics take place over a wide range of length scales (in the range of a few nm) and timescales (from milliseconds to picoseconds) influencing several aspects of protein function, no single technique can span so many spatial and temporal orders of magnitude. Therefore, there has obviously been a debate about how to reconcile the results of different experimental methods that explore dynamics [1, 2, 22, 40, 41].

Recent realization that low-frequency and large-amplitude modes of water molecules in the hydration shell are particularly important in controlling the

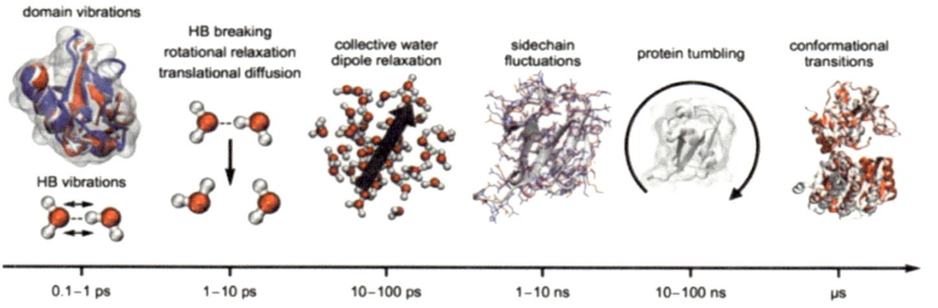

Figure 1.
The hierarchy of time scales for motions of proteins and their hydration environment. (Adopted with permission from ref. [43], Copyright (2015) American Institute of Physics).

conformational changes that dominate protein function, has led to using the terahertz (THz) spectroscopy to observe water dynamics around biological molecules [42, 43]. THz radiation (1 THz = 10^{12} Hz = 33.33 cm^{-1}) excites the low frequency vibrations of the solvated protein and directly probes the collective hydrogen bond dynamics of the coupled protein-water system in (sub-) ps time domain. Librational, translational and intermolecular, collective motions of hydration water, as well as large amplitude motions of biomolecules, also occur in similar time scale (see **Figure 1**). In frequency domain, this corresponds to low-frequency modes in the 1–10 THz frequency region. In spite of the fact that water molecules strongly attenuate the THz radiation in this frequency region, laser or accelerator-based THz sources (vide Section 2) are now powerful enough to penetrate water layers. Modern THz spectroscopic techniques can provide valuable information about water dynamics in the frequency region of the electromagnetic spectrum (0.3–20 THz or 10 to 600 cm^{-1}, the so-called "terahertz gap" between the dielectric and the infrared regimes). This has offered a unique view of the hydration water in fully solvated biomolecules.

2. THz transmission or absorption spectroscopy

Two kinds of THz spectroscopy techniques, namely THz transmission and THz time-domain spectroscopy (THz-TDS), have extensively been used to record the THz absorption spectra of biological materials. THz transmission spectroscopy is a simple single or dual beam steady state spectrometer using THz radiation beam to estimate the THz absorption coefficient. THz transmission or absorption spectroscopy of fully solvated biomolecules in water yields direct information on the global dynamical correlations among solvent (water) molecules. However, application of THz spectroscopy to study solvated biomolecules was not possible because of huge absorption of water in the THz frequency range. This problem was solved by the development of the p-Ge laser, which was a strong the THz emitter [44–47]. Prior to this development, weak radiation power of standard sources like a globar or an arc lamp in the far IR and THz region was the cause for poor signal to noise ratio and the measurements were limited to powders or hydrated films of biomolecules, but not in their native aqueous environment. Free-electron lasers [48, 49] and synchrotrons [50], which are also sources of high power far-IR and THz pulses, have also been reported to be used for spectroscopy studies of biomolecules but are not easily accessible. THz transmission spectroscopy of water in biological systems received a momentum after the discovery of the p-Ge laser. This laser is a powerful table-top THz source (up to about 400 W) and has the tunability in the 80–100 cm^{-1} (2.4-3.3 THz) frequency region. Development of a table top dual beam (reference and sample arms) THz transmission spectrometer ensured accurate measurements of absorption of water of thickness as large as 100 μm [51].

Accelerators delivering ultrashort pulses of relativistic electrons have been widely used as a source of high intensity THz radiation. Free electron lasers and synchrotrons are actually seeded by short duration electron pulses. An easier method of generating high intensity and tuneable THz radiation using an ultrafast electron accelerator is the coherent transition radiation (CTR). CTR occurs when a charged particle passes through an interface between media with different dielectric constants [52–55]. Sudden change in the dielectric constants along the electron's path causes a discontinuity in the electric field at the interface and this discontinuity readjusts itself as radiation spreading out from the point where the electron passes through the discontinuity. The angular spectral energy density of the CTR depends on the dielectric constants of the two media [56, 57]. Since the dielectric constants of aluminum and gold are much more than that of vacuum in far-IR region, these two metals are frequently used as perfect conductors. Direction of propagation and the intensity distribution of the CTR with respect to the angle of incidence of the electron pulse on the target has been discussed in detail in Ref. [53].

At wavelengths shorter than the bunch length, the emitted radiation field is incoherent and total intensity is proportional to the number of electrons (N). But at wavelengths longer than the electron bunch length, the radiation emitted from the bunch is coherent. With a typical number of electrons per bunch on the order of 10^8–10^9, the coherent radiation intensity greatly exceeds that of incoherent radiation. Especially, at long wavelengths compared to the bunch length, the radiation intensity is proportional to the square of the number of electrons in the bunch. Therefore, it is possible to generate coherent radiation in the far-IR or THz spectral range from short electron bunches with bunch lengths of hundred micron or less. Moreover, the shorter the bunch length, the broader is the radiation spectrum that can be generated. With larger number of electrons in the bunch, the total intensity (both coherent and incoherent components) of the CTR at frequency ω is given by,

$$I_{total}(\omega) = NI_e(\omega)(N-1)[1+(N-1)f(\omega)] \tag{1}$$

In this equation, contribution of the coherent component, $I_{coherent}(\omega)$, is given by,

$$I_{coherent}(\omega) = N(N-1)I_e(\omega)f(\omega) \tag{2}$$

Here,

$$f(\omega) = e^{-\left(\frac{\omega\sigma_z}{c}\right)^2} \tag{3}$$

and $I_e(\omega)$ is the radiation intensity generated from one electron in the bunch and Eqs. (1) and (2) have been derived assuming that all electrons in the bunch have the same energy. $f(\omega)$ is the bunch form factor at frequency ω, and it is the absolute square of Fourier transformation of the normalized bunch distribution. σ_z is the Gaussian bunch width in μm. Moreover, the shorter the bunch length, the broader is the radiation spectrum that can be generated. The Gaussian electron bunches with bunch lengths of 100, 200 and 300 fs can provide broadband radiation spectra covering the wavenumbers up to about 220, 110 and 70 cm^{-1} (or 6.6, 3.33 and 2.1 THz), respectively [55].

In THz transmission measurements at AIST, Tsukuba, Japan, a femtosecond linear accelerator system, which delivered electron pulses of about 300 fs duration, and electron energy of 42 MeV and ~ 100 pC charge (~10^8 electrons per one macro-pulse) [58], was used. Electron beam was directed to hit a thin gold foil (thickness is about 500 μm) at an incidence angle of 45°. CTR, which contained the component of THz radiation, was collected in the direction perpendicular to the electron beam using a gold coated parabolic mirror and the intensity of the THz radiation was measured using a Schottky diode detector [54]. Intensity of the THz pulse generated by this method was of high intensity (about 100 nJ/micropulse and covered a wide frequency range (0.3–2.5 THz), which was practically determined by the frequency response of the Schottky diode detector. Band pass filters (Tydex, Russia) were used to select a band of THz frequencies for estimation of absorption of protein solutions. Absorbance of the solutions of different path lengths were measured. Samples were taken in Bruker liquid sample cells, path lengths of which were varied using Teflon spacers of required thicknesses. Fused silica windows having thickness of 1.5 mm were used in the Bruker sample cell. Thicknesses of the Teflon spacers used were in the range of 20 to 150 μm. Absorption coefficient value (in cm^{-1}) of the solution was estimated from the slope of the linear plot of absorbance vs. path length of the cell (Eq. (4)) [59]. Experimental arrangement used here corresponded to that of a single beam absorption spectrometer.

$$ln\left(\frac{I_0}{I_d}\right) = \alpha_{sol}d \tag{4}$$

Here, I_0 and I_d represent the beam intensities in the absence (i.e. using the blank cell) and presence of the sample solution, respectively. α_{sol} represents the absorption coefficient of the sample solution, d is the path length or thickness of the sample.

The absorption coefficient of the sample, α, is defined by,

$$\alpha = \frac{4\pi\overline{\nu}k}{c} \tag{5}$$

$\overline{\nu}$ (cm^{-1}) is the frequency of THz radiation, k is the imaginary part of the complex refractive index and c is the speed of light. The complex refractive index is defined as,

$$\overline{n}(\overline{\nu}) = n(\overline{\nu}) + i\,k(\overline{\nu}) \tag{6}$$

The real part $n(\overline{\nu})$ describes the refractive index of the sample, while the imaginary part k is the absorption.

Presently, the THz-TDS in the 0.1–10 THz (3.33 cm^{-1}– 333 cm^{-1}) region is the most popular technique, which has been extensively applied in the fields of research in chemistry, materials science, physics, engineering, medicine as well as in industry [60–74]. THz -TDS has also been proved to be a valuable technique for investigation of low frequency dielectric relaxation and vibrational spectroscopy of hydrogen bonded liquids, such as water, alcohols, and others [75–77]. While various configurations of the THz time-domain spectrometer have been used depending on the kind of applications, we will describe here the general principle of the technique and the spectrometer configuration used for investigation of biomolecules.

The core principle of the THz-TDS technique, which uses a short duration (say a few tens of femtosecond) pulses of THz radiation, is measurement of the transient electric field associated with the THz pulse rather than variation of intensity of the frequency components of the THz pulse. However, since the response or the rise times of the electronic components and detectors are slower than a few ps, it is the normal practice to use the pump-and-probe kind of optical configurations in ultrafast spectrometers in order to overcome the slow response of the detectors. To achieve sub-picosecond time resolution in THz-TDS, the ultrashort NIR optical pulse (typically shorter than 100 fs) is beam-split along two paths to generate pump and probe pulses and to detect the time dependent THz field using a time-delay stage. Schematic diagram presented in **Figure 2** represents the principle of a THz-time-domain spectrometer.

The pump pulse, which contains the major amount of energy (more than 70% of the total energy of the laser beam), is used for generating broadband pulses of THz radiation using a photoconductive antenna consisting of a low temperature grown GaAs or InGaAs film covered with metallic contacts for application of bias voltage, or by optical rectification in ZnTe(110) crystal. The pulsed beam of THz radiation is collimated and focused onto the sample using gold coated parabolic mirrors. After transmission through the sample, the THz beam is recollimated and refocused on another photoconductive antenna or ZnTe(110) crystal, which works as the THz detector. Recently, ZnTe (110) crystal are being used extensively both as THz emitter as well as in detection because of possibility of its application in much wider THz frequency region as compared to that has been possible using photoconductive antenna.

The probe beam is used to gate the detector and measure the instantaneous THz electric field using the method of electro-optic sampling. A high precision linear motion stage is used to delay the probe pulses to arrive at detector crystal with respect to the pump pulses. The THz pulse and the gate (NIR) pulse are propagated collinearly through the ZnTe crystal. The THz pulse induces a birefringence in ZnTe crystal, which is read out by a linearly polarized gate pulse. When both the gate pulse and the THz pulse are present in the crystal at the same time, the polarization of the gate pulse will be rotated by the THz pulse depending on the strength of the electric field of the THz pulse. Using a $\lambda/4$ waveplate and a beam splitting polarizer together with a set of balanced photodiodes, the THz pulse amplitude is mapped by monitoring the gate pulse polarization rotation after the ZnTe crystal at various delay times with respect to the THz pulse (**Figure 2(a)**). A Fourier transform is then used to convert this time domain electrical signal to a frequency domain spectrum (**Figure 2(b)**). The optical path from THz generation to THz detection is purged with dry nitrogen gas to avoid the effect of the humidity.

Dielectric relaxation measurement is a sound method to investigate intermolecular interactions and is capable of monitoring cooperative processes at the molecular level [78–80]. This method is appropriate to monitor molecular motions in widely varied time scales and has been used extensively to study the structure, dynamics, and macroscopic behavior of complex systems [81]. A brief

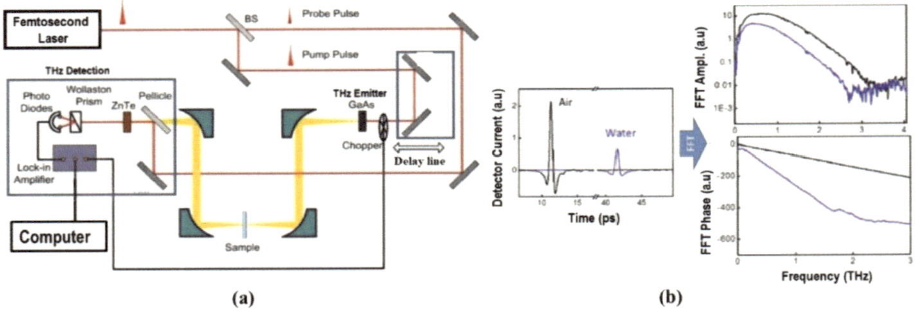

Figure 2.
(a): Schematic diagram of a typical THz - time-domain spectrometer. (b): Temporal variation of electric field of the THz pulse and its Fourier transform in air and in presence of water.

discussion (more details may be found elsewhere [81]) on the basic principle of dielectric polarization is presented below.

Dielectric polarization in static electric field: If a dielectric material is placed in an external static electric field $\vec{E}$, there occurs a charge displacement, which in turn creates a *macroscopic dipole moment* ($\vec{M}$) and this process is known as polarization ($\vec{P}$) or dipole density (**Figure 3**). This polarization is related to the $<\vec{M}>$ (ensemble average) and volume (V) of the sample by the following equation,

$$\vec{P} = \frac{<\vec{M}>}{V} \tag{7}$$

If a static electric field of strength, $\vec{E}$, is applied to an isotropic and uniform dielectric material of dielectric susceptibility χ, then the macroscopic polarization can be expressed as,

$$\vec{P} = \varepsilon_0 \chi \vec{E} \tag{8}$$

where ε_0 is the dielectric permittivity of the vacuum. Applying the macroscopic Maxwell theory, the electric displacement (electric induction) vector, $\vec{D}$ can be written as

$$\vec{D} = \varepsilon_0 \vec{E} + \vec{P} = \varepsilon_0 (1 + \chi) \vec{E} \tag{9}$$

Now, in the linear regime, the relative dielectric permittivity (or *dielectric constant*) is independent of the field strength, $\varepsilon = 1 + \chi$. Therefore,

$$\vec{D} = \varepsilon_0 \varepsilon \vec{E} \tag{10}$$

There are different types of polarization mechanisms by an applied electric field in different frequency regions: i) *Ionic polarization*, ii) *Orientation or dipolar rotation polarization*, iii) *Deformation (Atomic and Electronic) polarization*. The total polarization ($\vec{P}$) is the sum of *induced polarization* ($\vec{P}_\alpha$) and *dipole polarization* ($\vec{P}_\mu$).

Therefore,

$$\vec{P}_\alpha + \vec{P}_\mu = \varepsilon_0 (\varepsilon - 1) \vec{E} \tag{11}$$

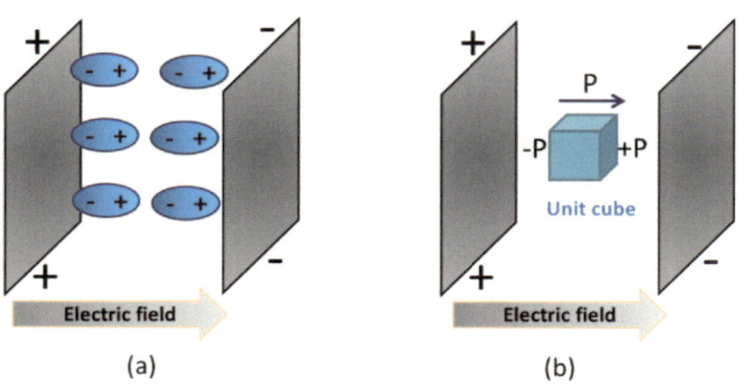

Figure 3.
(a) Polarization of dielectric. (b) Polarization vector.

For *polar dielectric,* the individual molecule possesses a permanent dipole moment even in the absence of any external electric field but for the *nonpolar* one, there is no dipole moment unless an electric field is applied. Due to the long range of the dipolar forces, an accurate calculation of the interaction of a particular dipole with all other dipoles of a specimen would be very complicated. However, a good approximation can be made by considering that the dipoles beyond a certain distance can be replaced by a continuous medium having the macroscopic dielectric properties. Assuming a continuum with dielectric constant ε_∞, in which point dipoles with a moment, μ_d, are embedded, we can write the induced polarization as follows,

$$\vec{P}_\alpha = \varepsilon_0(\varepsilon_\infty - 1)\vec{E} \tag{12}$$

The orientation polarization is given by the dipole density due to the dipoles $\vec{\mu}_d$. If we consider a sphere with volume V comprised of dipoles, we can write,

$$\vec{P}_\mu = \frac{<\vec{M}_d>}{V} = \frac{\sum_i (\vec{\mu}_d)_i}{V} \tag{13}$$

Dielectric polarization in time-dependent electric field: When a time- dependent electric field is applied, the decay function of dielectric polarization is given by,

$$\phi(t) = \frac{\vec{P}(t)}{\vec{P}(0)} \tag{14}$$

where $\vec{P}(t)$ is the time-dependent polarization vector. Time-dependent displacement vector $\vec{D}(t)$ can be written as follows,

$$\vec{D}(t) = \varepsilon_0\vec{E}(t) + \vec{P}(t) \tag{15}$$

$$\vec{D}(t) = \varepsilon_0\left[\varepsilon_\infty\vec{E}(t) + \int_{-\infty}^{t} \Phi(t')\vec{E}(t-t')dt'\right] \tag{16}$$

where $\Phi(t)$ is the dielectric response function, $\Phi(t) = (\varepsilon_s - \varepsilon_\infty)/[1 - \varphi(t)]$. ε_s and ε_∞ are the dielectric permittivity at low and high frequency, respectively. The frequency dependent complex permittivity $\varepsilon^*(\omega)$ is connected to the above relaxation function through Laplace transformation [82–84],

$$\frac{\varepsilon^*(\omega) - \varepsilon_\infty}{\varepsilon_s - \varepsilon_\infty} = \hat{L}\left[-\frac{d}{dt}\varphi(t)\right] \tag{17}$$

where $\hat{L}$ is the operator of the Laplace transformation, which is defined for the arbitrary time dependent function $f(t)$ as,

$$\hat{L}[f(t)] \equiv F(\omega) = \int_0^\infty e^{-pt} f(t)dt \tag{18}$$

where $p = x + i\omega$ and $x \to 0$.

Eq. (17) gives equivalent information on dielectric relaxation properties of the sample both in frequency domain and time domain measurements. Therefore, the dielectric response might be measured experimentally as a function of either frequency or time, providing data in the form of a dielectric spectrum $\varepsilon^*(\omega)$ or the macroscopic relaxation function $\varphi(t)$. For instance, when a macroscopic relaxation function obeys the simple exponential law,

$$\varphi(t) = \exp\left(-\frac{t}{\tau_m}\right) \tag{19}$$

Where, τ_m indicates the relaxation time. Eqs. (17) and (19) give rise to the following relation,

$$\frac{\varepsilon^*(\omega) - \varepsilon_\infty}{\varepsilon_s - \varepsilon_\infty} = \frac{1}{1 + i\omega\tau_m} \tag{20}$$

i.e,

$$\varepsilon^*(\omega) = \varepsilon_\infty + \frac{\varepsilon_s - \varepsilon_\infty}{1 + i\omega\tau_m} \tag{21}$$

This is known as the Debye model for frequency dependent dielectric permittivity. According to the Debye model, the complex frequency dependent dielectric response $[\tilde{\varepsilon}(\nu) = \varepsilon_{real}(\nu) - i\varepsilon_{imaginary}(\nu)]$ can be described as:

$$\tilde{\varepsilon}(\nu) = \varepsilon_\infty + \sum_{j=1}^{m} \frac{\varepsilon_j - \varepsilon_{j+1}}{1 + i2\pi\nu\tau_j} + \frac{\sigma}{i2\pi\nu\varepsilon_0} \tag{22}$$

where, ε_0 is the permittivity in free space (= 8.854 x 10^{-12} F/m), $\omega = 2\pi\nu$ is the angular frequency, τ_j is the relaxation time for the j-th relaxation mode, ε_1 is the static dielectric constant, ε_j are the dielectric constants for different relaxation processes, ε_∞ is the extrapolated dielectric constant at a very high frequency and m describes the number of relaxation modes.

3. Structure and dynamics of hydration layer in interfaces of small molecules, proteins, peptides, nucleic acids and membranes

The dynamics of water around small as well as complex molecules changes owing to their specific interaction with the solute surface; the specific nature of the interaction could mostly be electrostatic or hydrogen bonding. Such interaction expectedly ruptures/modifies the tetrahedral water network structure as well as its dynamics, a clear imprint of that gets reflected in the THz frequency window. Measurement of optical parameters $\alpha(\nu)$ and $n(\nu)$ enables one to determine the change in water dynamics at the vicinity of (bio)surface and consequently one can infer on the changes of the corresponding solute also.

3.1 Ions

Metal ions are perhaps the simplest solutes that can induce perturbation in water dynamics, the interaction being mostly electrostatic in nature. A systematic investigation using alkali monovalent cations [85] and alkaline earth bivalent cations [86] in aqueous solutions have been put forward by the group of Martina Havenith using far-IR and THz FTIR measurements coupled with classical MD simulation studies.

Their study concluded that ion rattling motions can account for the observed changes in the THz absorption. They have shown that the spectrum of the salt solutions can be approximated by a linear superposition of concentration weighted neat water and ion contributions. Ab initio MD simulation study by Marx et al. [87] has successfully reproduced the spectral responses of the solvation shell around the ions in infrared and THz frequency range. Their study has shown that the solute-solvent dipolar couplings and the dipole–dipole correlations are the important factors that govern the absorption features in this frequency regions. The group of Martina Havenith has subsequently put forward a series of studies on the hydration dynamics of heavy metal ions, e.g. lithium [88], manganese [89], iron [90] and ytterbium [91]. An elaborate description of the process involved has recently been put forward by this group [92]. The overall THz effect of the ions could be apprehended in terms of the contributions from various modes as depicted in **Figure 4**.

In a systematic study using THz-TDS in the frequency region (0.3–2.1 THz; 10–70 cm^{-1}), experimental evidences for the ultrafast collective hydrogen bond dynamics of water in the extended hydration layers of alkali metal chlorides have been obtained [93]. The real and imaginary part of the permittivity (ε), as obtained from the THz-TDS measurements was fitted using a triple Debye relaxation model. The time scales obtained for bulk water are of the order of ~8–9 ps (τ_1), 200 fs (τ_2) and 80 fs (τ_3). The ~9 ps and ~ 200 fs timescales are due to the well-known cooperative rearrangement of the H-bonded network structure and the small angular rotational modes of individual polar water molecules, respectively. It has been reported that (τ_1) decreases with increasing salt concentration, which identifies an acceleration of the cooperative hydrogen bond dynamics affirming a positive support towards the most debated notion of these ions to act as water structure breakers (see **Figure 5(a)**). The extent of this effect has been found to be mostly ion specific, K^+ being the most effective ion and a simple consideration of ionic charge density is insufficient to account for the observed changes. Very recently, Havenith and Marx [94] have put forward a combined experimental and simulation investigation to provide a detailed mechanistic analyses based on cross-correlation analysis (CCA) technique to understand the minute details of water-solute interactions.

3.2 Complex ions and small molecules

While the interaction of metal ions with water is more straight-forward, complexity arises when an ion is associated with a hydrophobic moiety. Such molecules

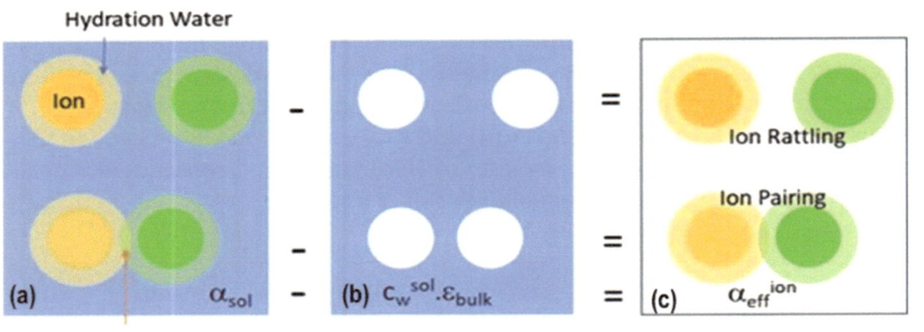

Figure 4.
Schematic representation of ion hydration. (a): Bulk-like water (blue), water around ions (yellow and green), and hydration water (lighter shades of yellow and green). The total absorption of the solution (α_{sol}) has contribution from all. (b): Concentration-weighted bulk-water values are subtracted from αsol. (c): This yields the effective absorption of the solvated ion or ion pair (α_{eff}^{ion}).

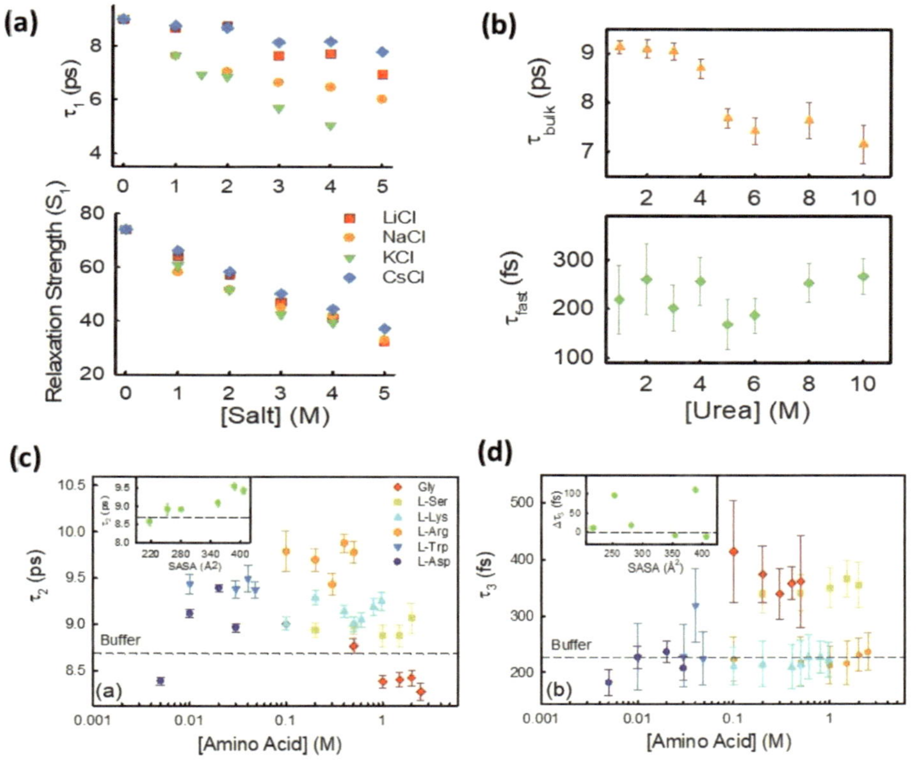

Figure 5.
(a) Cooperative relaxation dynamics (τ_1) and relaxation strength (S_1) of aqueous solutions of different alkaline metal cations. (b) Relaxation dynamics (as defined by the timescales τ_1 and τ_2 of aqueous urea solution as a function of urea concentration. A sharp change in the τ_1 is observed at ~4 M urea. (c) Cooperative hydrogen bond relaxation time constant (τ_2) of aqueous solutions of amino acids as a function of their concentration. The dotted line is the time constant for the buffer solution. The inset shows the change in τ_2 (measured at maximum amino acid concentration) as a function of the SASA of amino acid. (d) Relaxation time scale τ_3 corresponding to the jump orientation of water (obtained by Debye fitting of THz data) of different amino acids solutions. The dotted line is the time constant for that of the buffer. The inset shows the change in the timescale ($\Delta\tau_3 = \tau_{solution} - \tau_{buffer}$) at the maximum concentration of the amino acids.

are often of biological importance and therefore their interaction with water is essential to establish. There have been several reports on the THz studies of sucrose and saccharides [95–99]. In a pioneering work Heyden et al. [100] have described the nature of hydration around small carbohydrate molecules and a detailed simulation study has revealed that the extent of the hydration shell and their absorption co-efficient change with the solute specificity. One such molecule is urea, which is well known to be a protein denaturant, however, the exact molecular mechanism of the processes involved still remains a strongly debatable issue, specially the argument of whether the interaction is water mediated (the water structure breakers notion of urea) or a direct interaction of urea with protein surface.

An elaborate attempt has been made to understand the effect of urea (and its derivatives) on the ultrafast solvation dynamics of water using a Debye type dielectric relaxation model in the 0.3–2.0 THz frequency region using THz-TDS technique [101]. The relaxation dynamics shows considerable acceleration beyond a threshold concentration. Such acceleration is possibly associated with a disruption of the tetrahedral water network structure. It seems also intriguing that the observed collapse occurs at a certain urea concentration (see **Figure 5(b)**), which strikingly coincides with the denaturation concentration of urea for many proteins.

Another molecule of such biological interest is guanidinium chloride (GdmCl), which offers remarkable protein denaturation ability beyond a certain threshold concentration (~2–3 M) [102, 103]. Like in the case of urea, protein denaturation mechanism of GdmCl has also been debatable concerning a direct or indirect interaction. We have investigated the collective hydrogen bond dynamics around GdmCl in aqueous solutions as well as in the presence of a model globular protein human serum albumin (HSA) using the THz-TDS technique. It is found that the relaxation dynamics gets faster which renders support to the previously speculated notion that GdmCl acts as a water structure breaker. A similar but more prominent trend is observed in case of NaCl, which, however, does not interact with proteins.

The change in hydration dynamics in presence of HSA has been found to be in complete contrast in these two salts unambiguously pointing out towards a possibility of the collective hydration dynamics to share a pivotal contribution in the protein unfolding phenomenon. These studies seem to conjecture that ions with complex hydrophobic moiety acts as water structure breaker, and this very nature of such molecules makes them protein denaturant. To investigate the validity of such conjecture we investigated the ultrafast (sub-ps to ps) collective hydrogen bond dynamics of water in the extended hydration layers in a series of alkylammonium chloride salts using THz -TDS technique [104]. We found these salts to transform from being a water 'structure breaker' to 'structure maker' with increasing carbon content. For example, the THz-TDS measurements reveal that ammonium chloride (AC) acts as a water network structure breaker while tri-ethyl ammonium chloride (EAC) and higher carbon containing salts are distinct water structure makers. The change in protein hydration is also found to trace its secondary structure rupture and the exposure of hydrophobic moieties accordingly changes the protein hydration. Our study strongly concludes that it is the hydrophobic effect, at least in the case of this type of salts, that plays the decisive role in determining their interaction with biomolecules.

3.3 Amino acids

Amino acids, which are the building blocks of proteins, often act as intermediates in metabolism as well as osmolytes that can stabilize proteins. Depending upon the 'side group', the amino acids are classified as hydrophilic or hydrophobic. Hydrophobicity of amino acids is believed to be a key parameter that regulates phenomena like protein folding - unfolding, aggregation, activity, protein-ligand binding and protein hydration in aqueous environments. While to analyze the solvation of hydrophobic and hydrophilic parts of a protein separately, one needs to take into consideration that the environment of amino acid residues is heterogeneous in nature as they are often composed of hydrophobic alkyl chains and hydrophilic groups. It is therefore essential to study the hydration of amino acids of various side chains in the exposure to solvents. Niehues et al. [105] have studied the hydration of a series of amino acids in ~2.4 THz window and observed that the THz absorption coefficient, $\alpha(\nu_{THz})$, of hydrated amino acids can be correlated with the hydrophobicity and the fraction of polar volume of amino acids. They have shown that glycine has the largest positive THz slope followed by serine, whereas for the other amino acids, the slope becomes gradually negative with increasing hydrophobicity. In another study [106], the same group, by analyzing the corresponding THz spectrum in terms of the correlated dynamics of solute and solvent molecules, demonstrates the line shape of the low-frequency vibrational response of glycine in water. Recently, they [107] have shown that hydrophilic solvation of the zwitterionic groups in valine and glycine leads to similar THz responses, which are fully decoupled from the side chain. This result concludes that the hydrophilic groups and their solvation shells dominate the THz absorption difference, while on the same intensity scale, the influence of hydrophobic water can be neglected. Shiraga et al. [108] concludes a correlation between the extent of H-bonding and the hydrophobicity of the solute. Glycine (Gly) and the L- isomers of five different amino acids: serine (Ser), aspartic acid (Asp), lysine (Lys), arginine (Arg) and tryptophan (Trp) of varying hydrophobicity and solvent accessible surface area (SASA) have been used by Samanta et al. [109] to study the dielectric relaxation up to their maximum water solubility in the frequency window of GHz-THz at neutral pH. The various rotational dynamics of the solutes and water are obtained by fitting the dielectric data in a multiple Debye relaxation model. From GHz study, we observed that the molecular rotation of amino acids correlates their respective molecular volume. Also, it was concluded that the amino acids do not usually aggregate even at high concentrations. From THz study, the authors found that Gly is a water structure breaker while the other amino acids are structure makers. Consequently, Gly accelerates cooperative hydration, while the others retard (see **Figure 5(c)** and **(d)**). It has been established that hydration in amino acids depends both on its hydrophobic as well as its hydrophilic nature of side chains. Born et al. [110] have studied the hydration dynamics of some small model peptides using THz spectroscopy. They observed that the cooperative hydration dynamics changes as the extent of hydration of such molecules change.

3.4 Alcohols and other solvents

Addition of otherwise indifferent organic solvents can influence marked changes in aqueous environments as well as they can induce misfunction in biologically important molecules e.g. protein, DNA, etc. Water - dimethyl sulfoxide (DMSO) binary mixture has been found to be of potential interest as it plays a significant role in the field of chemistry, physics, biology and pharmacology. Das Mahanta et al. have investigated the change in the collective hydration dynamics in presence of DMSO at different concentrations using THz-TDS. We found that $\alpha(\nu)$ of the mixed solvents shows a non-linear change with X_{DMSO} [111]. This change has been correlated with the water-DMSO structural heterogeneity in the mixed solvents. Luong et al. have also studied the evolution of water hydrogen bonded collective network dynamics in water −1,4-dioxane (Dx) mixtures as the mole fraction of water (X_w) increases from 0.005 to 0.54 [112]. The inter- and intra-molecular vibrations of water be observed using THz-TDS in the frequency range 0.4–1.4 THz (13–47 cm^{-1}) and Fourier transform infrared (FTIR) spectroscopy in the far-infrared (30–650 cm^{-1}) regions. From the absorption coefficient measurements, they infer that the mixtures are not ideal in nature, which suggests a significant change in the network by the addition of the solute. The authors found an increase in the collective hydrogen bond network as evidenced from dielectric relaxation studies in which τ_1 has been found to be small at low X_w (where X_w is the mole fraction of water in the mixture), but at $X_w > 0.1$, it increases rapidly to reach a value identical to that in bulk water. It has been concluded that hetero-molecular (water - Dx) hydrogen bond dominates in the water diluted region in water-Dx mixtures, and with progressive addition of water, bulk-like intermolecular three-dimensional hydrogen bonded water network dynamics evolves beyond $X_w = 0.1$. In a separate study [113], a combined experimental (mid- and far-infrared FTIR spectroscopy, THz-TDS (0.3–1.6 THz)) and molecular dynamics (MD) simulation technique has been carried out to understand the evolution of the structure and dynamics of water in its binary mixture with 1,2-dimethoxy ethane (DME) over the entire concentration range. Debye relaxation data reveals a non-monotonous behavior in which the collective dynamics is much faster in the low X_w region, whereas in the $X_w \sim 0.8$

region, the dynamics gets slower than that of pure water. The concentration dependence of the reorientation times of water, estimated from the MD simulations, also captures this non-monotonous character. Bohm et al. [114] have demonstrated that the THz response of alcohols can be decomposed into the spectrum of bulk water, tetrahedral hydration water, and more disordered (or interstitial) hydration water. They also concluded that it is not the tetrahedrally ordered component, rather it is the interstitial hydration water which is responsible for the temperature-dependent change in ΔC_p and ΔG in such mixtures.

3.5 Proteins

The surface of proteins is extremely heterogeneous owing to the presence of amino acids of varying types of charges. The pioneering studies by the group of Havenith et al. have shown that protein molecules are hydrated and the cooperative dynamics of water changes accordingly. An earlier study using a small protein ubiquitin [115] shows how the measurement of $\alpha(\nu_{THz})$ reveals a change in the protein hydration dynamics as the authors termed it as the "THz dance". In a subsequent ever important simulation paper, the same group has established different THz absorption of the hydrophobic and hydrophilic residues of a protein as they interact differently with water [116]. In the later years, this group has put forward substantial contribution on the hydration dynamics of proteins using THz measurements [117, 118] as well as simulation studies [42, 119]. Their studies have unambiguously suggested that fast protein motion and solvent dynamics are correlated with enzymatic reactions [120]. In a recent study this group has established the pivotal role of collective motion of water during an electron transfer reaction between flavoenzyme ferredoxin - $NADP^+$ – reductase and ferredoxin-1 [121]. In a seminal paper, Marklez et al. [122] have used THz-TDS measurements to show that the protein (hen egg white lysozyme) dynamical transition (the rapid increase in protein dynamics occurring at ~200 K) needs neither tertiary nor secondary structure. Their results revealed that the temperature dependency essentially arises from the protein side-chain interaction with the solvent. He et al. [123] have investigated the presence of structural collective motions on a picosecond timescale for the heme protein, cytochrome C, as a function of oxidation and hydration, using THz-TDS and molecular dynamics simulations. Marklez group has developed a novel measurement technique (anisotropy THz microscopy) [124] wherein they were able to detect the long range protein vibration modes in chicken egg white lysozyme single crystals. They found the underdamped modes to exist for frequencies $>10\,cm^{-1}$. Such underdamped vibrational modes have also been identified using optical Kerr effect measurements in the THz frequency window [125]. In a recent study, Niessen et al. [126] have used anisotropy THz microscopy, which is found to be sensitive to inhibitor binding and unique vibrational spectra for several proteins and an RNA G-quadruplex. There have been other reports from several experimental groups: Sun et al. [127] have reported the application of a new machine learning methods for quantitative characterization of bovine serum albumin (BSA) deposited thin-films detected by THz-TDS. The group of Emma Pickwell-MacPherson [128] have used THz spectroscopy to study the hydration shell formation around H9 subtype influenza A virus's HA protein (H9 HA). They have also detected antigen binding of H9 HA with the broadly neutralizing monoclonal antibody. They observed a remarkable concentration dependent nonlinear response of the H9 HA, which reveals the formation process of the hydration shell around H9 HA molecules. The same group has also reported the dielectric properties of two different antibodies in water-glycerol mixtures using THz-TDS measurements [129]. Sun et al. [130] used THZ-TDS to investigate the molecular processes involved above and below the transition temperature (T_D) for GP2 peptide.

3.6 DNA

While much research has been focused on proteins, relatively less attention has been paid on the other biologically important molecule, DNA. There had been a few preliminary studies to understand the vibronic bands in the THz frequency region using single and double stranded DNAs and RNAs [131–133]. Arora et al. [134] have presented a label free quantitative detection method for DNA samples amplified by polymerase chain reaction (PCR) in aqueous medium using THz-TDS in the frequency range from 0.3 to 1.2 THz. Tang et al. [135] have recently investigated the feasibility of THz spectroscopy combined with microstructures for marker-free detection of DNA and oligonucleotides. Polley et al. have investigated the collective dynamics of two DNA molecules extracted from salmon sperm and calf thymus and observed that the dynamics did not differ much at the concentration range of the experiments [136].

3.7 Lipid membrane

There have been only limited studies on the THz studies on lipid membranes and/or vesicles. One of the preliminary results was due to Tielrooij et al. [137] who had studied the dielectric relaxation in mixed system of hydrated DOPC lipid bilayers in the THz frequency domain. They could identify three distinct water types: fast, bulk and irrotational. The relative content of those change with the extent of hydration. Later, Yamamoto et al. [138] have studied the temperature and hydration dependent low frequency spectra of lipid bilayers of 1,2-dimyristoyl-sn-glycero-3-phosphoryl-3′-rac-glycerol (DMPG) and 1,2-dimyristoyl-sn-glycero-3-phosphocholine (DMPC) using THz-TDS. They found that the THz absorption patterns reflect the lipid packing pattern in the bilayers. They subsequently extended their investigations towards purple membrane (PM, a complex of lipids and a membrane protein, bacteriorhodopsin) [139] and lipid bilayer of DMPC [140]. Pal et al. have recently studied the microstructure and collective dynamics of the membrane interfacial hydration shell in zwitterionic and negatively charged phospholipid membrane bilayers using THz-TDS [141]. They observed a dependence of the critical lipid concentration corresponding to the inflection point on the charge of the lipid head-group, thereby implicating membrane electrostatics as a major factor in the microstructure and dynamics of water at the membrane interface.

4. Influence of temperature, pH, ionic strength and additives on hydration layer dynamics

Hydration dynamics of biomolecules is significantly influenced upon changing its physical conditions as well as in the presence of additional chemical agents, like alcohols, glycols, etc. Protein molecules undergo various physical and chemical changes, which could induce conformational modifications in their secondary as well as tertiary structures. It can be noted here that as protein structures get disrupted (during unfolding or denaturation) the hydrophobic moieties (amino acid residues), which are otherwise buried inside in the native structure, get exposed, and this produces a definite alteration in THz absorption coefficient, $\alpha(\nu_{THz})$. It therefore suggests that estimation of $\alpha(\nu_{THz})$ provides a direct evidence of the structural perturbation in proteins.

Several experimental techniques are available to determine the structural evolution of proteins during unfolding, while THz provides with the estimation of the associated hydration changes. In a pioneering experimental work, in which a stopped-flow techniques was synchronized with THz-TDS, which the authors termed as Kinetic terahertz absorption (KITA) spectroscopy, Kim et al. [142] have shown that the pH induced unfolding-refolding kinetics (in real time) of Ub* could easily be traced by the associated $\alpha(\nu_{THz})$ measurements. The THz results very well reconcile with the results obtained from circular dichroism (CD) and fluorescence measurements.

Another important physical environment that induces protein unfolding is temperature. In a study using HSA as a model protein, Mitra and Havenith [143] have shown that water dynamics associated with the protein during its reversible unfolding pathway up to 55°C as well as its irreversible denaturation pathway up to 70°C traces the protein's structural rupture pathway. The THz measurements do support the conventional CD and fluorescence measurements. Sudden increase in the environment (like temperature or pressure) for a very short period of time (often termed as T-jump experiments) leads protein molecules to be structurally ruptured but upon removal of the intense pulse, the protein refolds. T-jump experiments have previously been characterized using conventional CD and fluorescence measurements. However, THz measurements was demanded to obtain explicit information of water dynamics. The first report of such experiment was from the group of Havenith [144], wherein the authors put forward a coupled KITA setup with a T-jump attachment. The authors monitored changes in the THz absorption λ^{*6-85}protein with a time resolution of >50 μs. They reported that the spectral changes are correlated with the hydrophobic collapse of the protein. In a subsequent study from the same group, Wirtz et al. [145] used an even better time resolution of about 500 ns to reveal the coupled ubiquitin–solvent dynamics in the initial phase of hydrophobic collapse (temperature induced unfolding). They propose that, in the case of ubiquitin, a rapid (~500 ns) initial phase of the hydrophobic collapse from the elongated protein to a molten globule structure precedes secondary structure formation. Recently there have been a few reports of using THz spectroscopy technique to underline thermal denaturation of BSA [146], temperature- and pH-dependent protein conformational changes in pepsin A [147]. In a very recent study Cao et al. [148] have successfully employed THz-TDS to track the hydrolysis of BSA protein by pepsin. The results indicate that protein hydrolysis can be easily monitored over time by focusing on the variation of the absorption coefficient from a macroscopic perspective. The authors explored the use of the Debye model to analyze the dielectric properties of the solution during protein hydrolysis. The results of the Debye analysis prove that it is possible to investigate in detail the microscopic dynamics of bio-macromolecule solutions at the molecular level by THz-TDS.

Samanta et al. have been involved in determining the changes in protein hydration in various distressed environments using the THz-TDS measurements. They have investigated the hydration dynamics around HSA in presence of short chain polyethylene glycols of different chain lengths (PEG 200, PEG 400, and PEG 10000) at different concentrations [149]. FIR-FTIR studies conclude that the protein hydration is affected in a distinct way below and above the critical PEG concentration of 30% (v/v). THz-TDS study unambiguously confirmed a retardation of the solvation dynamics by PEGs. This study clearly shows an independent behavior of protein hydration at low PEG concentrations and a noticeable interaction between protein and PEG hydration beyond a critical PEG concentration. In another study, Das et al. have made an attempt to understand whether the DMSO induced conformation changes in lysozyme conformation perturbs its hydration dynamics [150]. CD study establishes a marked change in the protein tertiary structure in presence of DMSO. The relative change in the THz absorption coefficient ($\Delta\alpha/\alpha_0$) shows a negative minimum at X_{DMSO} = 0.05 and a positive value at

X_{DMSO} = 0.15. The observed minimum is found to be due to the increased size of the protein while the positive value is attributed to the increased SASA and consequent increased hydration of the protein surface. In a recent report, Das et al. put forward an experimental observation of nonmonotonic changes in the collective hydration of BSA in the presence of alcohols of varying carbon-chain lengths, that is, ethanol, 2-propanol, and tert-butyl alcohol (TBA), by using THz – TDS [151]. They observe an anomalous hydration behavior of the protein hydration with the alcohol concentration, which correlates the alcohol-induced α-helix to random coil transition of the protein secondary structure, as revealed by CD spectroscopy measurements. Recently, Das Mahanta have investigated the effect of alkyl-ammonium chloride salts on BSA and found a systematic trend towards disrupting the protein secondary structure [104]. The associated changes in the protein hydration in the presence of these salts have also been investigated using THz-TDS. The change in protein hydration is also found to trace its secondary structure rupture, and the exposure of hydrophobic moieties accordingly change the protein hydration. The THz-TDS measurements strongly conclude that it is the hydrophobic effect, at least in the case of this type of salts, that plays the decisive role in determining their interaction with biomolecules.

5. Role of hydration water in protein aggregation and fibrillation

Aggregated protein is toxic to functioning of living systems and many of human diseases are associated with misfolded protein disorders [152–156]. This is why understanding of mechanisms of interactions between protein molecules in solutions have been the subject of extensive investigations during the last two decades [157–161]. Additionally, understanding the protein aggregation propensity may offer novel design principles for producing aggregation-resistant proteins for biotherapeutics.

Globular proteins, e.g. HSA and BSA, in their native states are present in living cells at concentrations as high as about 200 mg mL^{-1} and bimolecular interactions are significant. We explained earlier that stability of the three - dimensional structure of a monomeric protein molecule in physiological environments is the result of an intricate interplay between electrostatic, hydrophobic, hydrogen bonding and other interactions and any variation of temperature, pH of the medium or any other physiochemical conditions including concentration of protein in the cell may result in imbalance of the stabilization forces leading to misfolding [162, 163], thus triggering aggregation [164–168].

A native and structurally stable protein molecule is strongly hydrated with a well-defined hydration layer of thickness of about a few tens of Å (20–40 Å) around it [50, 59, 169]. The role of water molecules in the hydration shell could be crucial for the intermolecular interactions and the overall protein hydrophobicity, which may be defined by its hydration free energy, which may play an important role in protein aggregation in aqueous solution [162, 163, 166]. However, the role of hydration water in protein aggregation has largely been unexplored owing to the perception that protein – protein interaction is the major factor and the surrounding water is just a spectator playing no role in aggregation of protein, ignoring water as an active constituent of biological systems. In fact, whether a protein remains soluble or forms aggregation should intrinsically rely on its state of hydration in the monomeric state.

We have described in the earlier sections of this article that how the properties of water molecules in the hydration layer on an average are different from those in the bulk, which determines the stability of a monomer protein in aqueous environment. Protein aggregation proceeds through a multistep process initiated by conformational transitions, called protein misfolding, of monomer species towards aggregation-prone structures. Chong and Ham applied the fluctuating thermodynamic analysis method to understand the variations of the thermodynamic functions, which occur during the course of misfolding and dimerization of the Amylase-β protein. They suggest that the time variation of the solvent-averaged effective energy, $F = E_u + G_{solv}$, describes the protein dynamics on the free energy landscape [170]. Here, the protein potential energy (E_u), comprises both intra and intermonomer contributions and the solvation free energy (G_{solv}), represents the interaction of the protein with surrounding water, which plays a critical role in protein aggregation. The free energy, F, decreases as the dimerization proceeds, but the decrease in F has different origins in the approach and structural adjustment regimes. The thermodynamic force driving the approach of two monomers is the decrease in G_{solv} and hence, the misfolded monomers acquire a large hydrophobicity, which leads to conformational changes in monomers. This drives two monomers to approach each other to a contact distance. In absence of this thermodynamic driving forces, two negatively charged protein monomers (the total charge of Amylase-β and BSA monomer proteins at neutral pH are −3, and − 16, respectively [170, 171]) would never approach each other by overcoming the electrostatic repulsion. On the other hand, decrease of the protein potential energy E_u, due to direct protein–protein interactions, such as intermonomer van der Waals contacts and hydrogen bonds, drives the structural rearrangement required for formation of compact dimer structure leading to energetic stabilization. Interestingly, structural rearrangements are also associated with an increase in the solvation free energy, which originates from the dehydration of the protein surface and of the interfacial region. On contrast to other spectroscopic techniques, THz spectroscopy probes directly the collective intermolecular vibrations of the hydrogen bond network, and is thus able to detect sensitively solute induced changes in the solvation dynamics. Extensive works on THz absorption of protein solutions have demonstrated that the absorption coefficient of the protein solutions (α_{sol}) are dependent on the concentration of protein [50, 59, 169–171]. For example, at the low concentration regime (e.g. in the case of HSA, $<0.5 \times 10^{-3}$ mol dm^{-3}) α_{sol} value increases linearly and this has been explained by increasing concentration of hydrated monomer protein molecules since water in the hydration shell has larger α_{sol} value as compared to that of bulk water. However, on further increase of the protein concentration, α_{sol} value starts decreasing.

To delineate this issue in more detail, Manna et al. made a detailed investigation on the concentration dependence of THz absorption of the aqueous buffered solutions of HSA protein (up to 2.6 mM of protein concentration) at three THz frequencies, namely, 0.1, 1.5 and 2 THz [59]. Similar results were obtained from these three measurements. α_{sol} value was expected to change linearly to follow Eqs. (23) and (24), which could be derived assuming that the protein solution is a two-component system.

$$\alpha_{sol} = \alpha_{pr} V_{pr} + \alpha_{bw} V_{bw} \tag{23}$$

$$\left(\frac{\alpha_{sol}}{\alpha_{bw}}\right) = \left(\frac{\alpha_{pr}}{\alpha_{bw}}\right) V_{pr} + V_{bw} = (1 - 0.08\,[HSA]) \tag{24}$$

Here, α_{pr} is the absorption coefficient of the protein, α_{bw} is the absorption coefficient of bulk water, V_{pr} is the volume fraction occupied by protein molecules and V_{bw} is the volume fraction occupied by bulk water. Eq. (24) was derived using several considerations, such as the value of the radius of gyration of HSA is about 3.3 nm [172], the structure of the monomer HSA protein molecule consists of hydrophobic and hydrophilic pores (2: 1 (v/v) ratio) with porosity factor of about 33% of the volume of the protein molecule [173].

Therefore, the plots of $\left(\frac{\alpha_{sol}}{\alpha_{bw}}\right)$ against HSA concentration, [HSA], as shown in **Figure 6** (red lines), represent the two-component system. However, a large deviation of the experimental data points from the line representing two component system suggests inadequacy of the two-component model. This deviation could be explained by considering that the THz absorption coefficients of the water molecules in the hydration layer, (α_{hl}), were different from that of bulk water (α_{bw}) and hence constitute the third component in the aqueous solution of the protein. Hence, Eq. (25), in which V_{hl} was the total volume fraction of water associated with the protein molecules in a solution with protein concentration of 1×10^{-3} mol dm^{-3}, was derived [59].

$$\left(\frac{\alpha_{sol}}{\alpha_{bw}}\right) = \left(\frac{\alpha_{hl} V_{hl}}{\alpha_{bw}}\right) C_{pr} + (1 - 0.08\, C_{pr} - V_{hl} C_{pr}) \tag{25}$$

To understand the reasons for significant decrease of THz absorption coefficient of the protein solutions with increasing concentration beyond 6 x 10^{-4} mol dm^{-3}, the possibility of aggregation of proteins at higher concentration regime was explored. To delineate this aspect, dynamic light scattering (DLS) measurements (**Figure 6**) were carried out using concentrations of proteins covering the entire range of THz absorption measurements. The DLS data recorded for the solution containing protein concentration of 0.4 x 10^{-3} mol dm^{-3} revealed the existence of only monomeric protein molecules with the most probable diameter of about 6–8 nm in the solution. This is quite in good agreement with the diameter of the

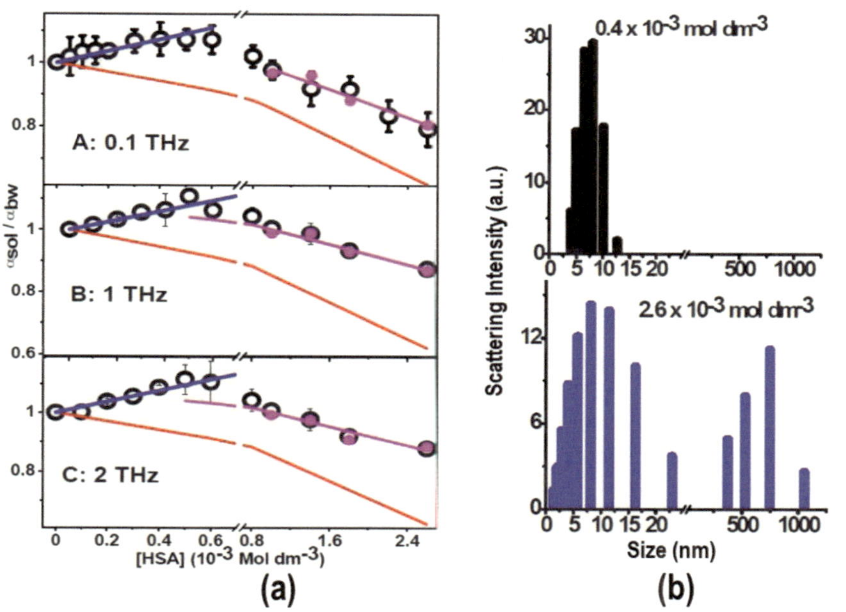

Figure 6.
(a): Plots of $\left(\frac{\alpha_{sol}}{\alpha_{bw}}\right)$ *vs. [HSA] recorded at three THz frequencies. Black squares with error bars are the experimental points. Red line represents two component system as per Eq. (24). Best nonlinear fit to the experimental points in the low concentration regime (up to 6 x* 10^{-4} *Mol dm*$^{-3}$*) using Eq. (25). Violet line is the linear fit to the experimental points at protein concentrations larger than 8 x* 10^{-4} *Mol dm*$^{-3}$ *using Eq. (26). (b): Size distribution of particles in solutions of two different CONCENTRATIONS of HSA as obtained from DLS analysis.*

hydrated monomer protein molecules. However, at higher concentrations (say, >1 x 10^{-3} mol dm^{-3} of protein), the DLS data revealed two important features. Firstly, the size distribution of the monomer band became wider indicating the presence of particles of diameter in the range 12–15 nm, possibly suggesting formation of dimers or trimers of HSA, in addition to the monomeric species. Secondly, at higher concentrations of the protein, DLS data also revealed the presence of large size aggregates of the most probable diameter of about 700 nm. However, a quantitative estimation of the relative percentages of monomer, dimer and aggregates was not possible from the DLS data because the intensity distribution of the scattered radiation was not directly proportional to the number of the particles. However, this experiment confirmed the presence of protein aggregates in solutions with higher concentrations of protein and the formation of aggregates may possibly be held responsible for nonlinear dependence of THz.

On the other hand, CD measurements confirmed that the tertiary structure of HSA remained unchanged through the entire range of HSA concentrations used for THz measurements. This suggested that the native structure of HSA molecules remained unaltered through the entire range of concentrations of HSA. Therefore, hydration states of proteins, even in the aggregated state, remain unchanged and hence possibly justifies the assumption made in the earlier works regarding overlapping of hydration shells at higher concentrations of proteins.

Patro and Przybycien have simulated the structures of reversible protein aggregates as a function of protein surface characteristics, protein–protein interaction energies and assessed the aggregate properties [174]. Results of their simulation reveals that aggregate particles have the kind of organization of the hydrophobic and hydrophilic domains as they are present in HSA protein monomer molecules and aggregation of protein molecules causes the loss in the total solvent accessible surface area (SAS) is about 67% and the mean solvent content for these aggregates vary in the range of 0.37–0.55 volume fraction depending on the conformation of the monomer protein [175]. Therefore, at higher concentrations of protein, volume fraction of hydration water decreases due to formation of aggregates. In addition, proteins are THz transparent and as we increase the protein concentration, protein aggregates replace the water molecules and leads to lowering of total THz absorbance of the solution.

A method of analysis was adopted to analytically fit the data in the regime of higher concentrations of the protein to predict the relative concentrations of the monomer and aggregated particles. In this analysis, the value of α_{hl}, which was estimated from the linear regime of the plot of $\left(\frac{\alpha_{sol}}{\alpha_{bw}}\right)$ vs. [HSA] was used. Assuming that the THz absorption coefficient of the water molecules residing inside the aggregate as well as that constituting the hydration layer around the surface of the aggregate are similar to that constituting the hydration layer around the monomer protein molecule, Eq. (9) was revised to write Eq. (11), which provided a quantitative estimate of the relative numbers of the HSA molecules in the monomeric and aggregated forms in solution.

$$\left(\frac{\alpha_{sol}}{\alpha_{bw}}\right) = \left(\frac{\alpha_{hl}}{\alpha_{bw}}\right) \left(V_{hl}^{ag} + v_{hl}x + (1 - 0.08\ V_{hl}^{ag} - Vv_{hl}x\right) \quad (26)$$

Here, x is the concentration of monomer in the unit of 1×10^{-3} mol dm^{-3} and V_{hl}^{ag} is the volume fraction of hydration water associated with the aggregates, both in the interstitial places as well as outside the aggregate constituting the hydration layer, v_{hl}, is the volume of hydration layer associated with one protein molecule.

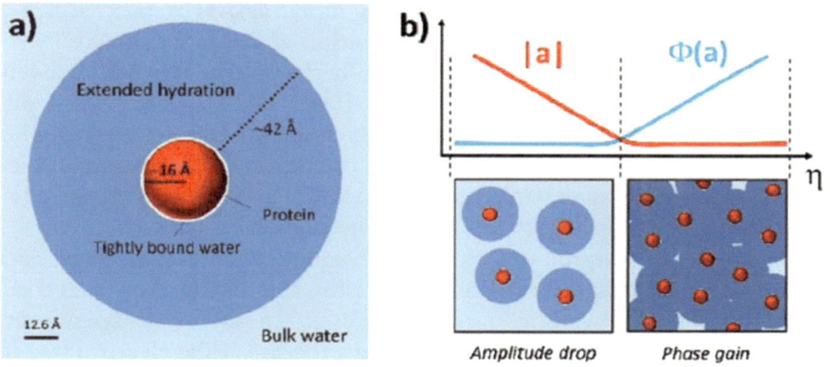

Figure 7.
(a) Cartoon of a human lysozyme protein (red sphere) in water. Water molecules tightly-bound to the protein surface (white), extended hydration layers (blue) and unperturbed bulk water (light blue). (b) Sketch of the evolution of the modulus and the phase of the induced dipole in a unit volume of solution versus protein concentration. The effect of phase gain at larger concentrations is represented by darker colors on the bottom right panel (adopted with permission from Ref. [170], Copyright (2017) American Chemical Society).

Manna et al. estimated the percentage of the number of HSA molecules, which exist as monomer (or dimer or trimer) in solution without being associated with aggregate formation, for each of the HSA concentrations used for THz absorption measurements [59]. We find that for the protein solution with its concentration of 8 x 10^{-4} mol dm^{-3}, about 40% of the protein molecules exists as monomer (or dimer or trimer) in solution and 60% protein molecules are part of aggregates. The monomer (or dimer or trimer) concentration becomes <1% at the protein concentration of 2.6 x 10^{-3} mol dm^{-3}, i.e. nearly all the protein molecules are associated with aggregate formation. These calculations suggest that formation of aggregates at higher concentrations of protein may be the responsible factor for the turnover of the TH_Z absorbance of the protein solution at ~6 x 10^{-4} mol dm^{-3} concentration.

The absorption coefficient of a complex system is the weighted sum of the absorption of its components. However, determination of the complex dielectric function of proteins in solution using THZ-TDS spectroscopy allowed detailed analysis beyond what was possible from simple absorption measurements. Following this approach, Novelli et al. determined both the phase and amplitude of the induced dipole in the volume containing the protein and hydration water [170]. This result revealed that not only the amplitude of the induced dipole varied with the concentration of protein, but also its phase changed above a concentration threshold (**Figure** 7). They proposed a phenomenological model, which explained that the phase of the induced dipole in the protein-solvent interaction region began to vary when there was significant overlap between hydration layers of the neighboring protein. This result suggested that indirect electromagnetic protein–protein interactions could take place if mediated by the extended hydration layers surrounding each protein.

6. Conclusions and challenges ahead

THz spectroscopy, from its very inception, has mostly been used by scientists studying cosmology, condensed matter physics and materials. The huge absorption of water in this frequency range used to be treated as a drawback of this technique; however, for chemists and biologists this point serves as an advantage since all the biological function is somehow or other related to water dynamics. Due to its inherent sensitivity to water hydrogen bonding dynamics, THz spectroscopy has become an indispensable tool for direct observation of fast and coupled biomolecule-water network. The initial studies by the groups of E. W. Heilweil, P.U. Jepsen, A. Marklez, C. Schmuttenmaer and M. Havenith in the late 1990s and early 2000s have established THz spectroscopy at a concrete platform to be recognized as a potential tool to label free detection of water dynamics in the vicinity of bio-molecules, the effect being extended to several layers and remains practically undetected using conventional spectroscopic methods. The last decade has witnessed a huge leap towards exploiting this frequency window in biophysical studies, a few of such results have been depicted in this article. Now that the phenomenon has been established beyond any doubt, new sort of experimental studies, where the role of hydration in ultrafast processes (like electron transfer or proton transfer) could explicitly be determined, is the new challenge to the researchers.

Author details

Rajib Kumar Mitra[1*] and Dipak Kumar Palit[2*]

1 Department of Chemical, Biological and Macromolecular Sciences, S.N. Bose National Centre for Basic Sciences, Kolkata, India

2 School of Chemical Sciences, UM-DAE Centre for Excellence in Basic Sciences, University of Mumbai, Mumbai, India

*Address all correspondence to: rajib@bose.res.in and dkpalit@cbs.ac.in

References

[1] Pal SK, Peon J, Bagchi B, Zewail AH. Biological water: Femtosecond dynamics of macromolecular hydration. J. Phys. Chem. B 2002;**106**:12376–12395

[2] Nandi N, Bhattacharyya K, Bagchi B. Dielectric relaxation and solvation dynamics of water in complex chemical and biological systems. Chem. Rev. 2000;**100**:2013–2045

[3] Ball P. Water as an Active Constituent in Cell Biology. Chem. Rev. 2008;**108**:74–108

[4] Wolynes PG. Recent successes of the energy landscape theory of protein folding and function. Quarterly Rev. Biophys. 2005;**38**:405–410

[5] Lynden-Bell R, Moris SC, Barrow JD, Finney JL, Harper JCL. *Water and Life: The Unique Properties of H2O*2010. pp. Boca Raton: CRC Press

[6] Bagchi B. *Water in Biological and Chemical Processes: From Structure and Dynamics to Function*2013. pp. Cambridge: Cambridge University Press:

[7] Hummer G, Tokmakoff A. Preface: Special Topic on Biological Water. J. Chem. Phys. 2014;**141**:22D101

[8] Goodsell DS. *The Machinery of Life*1993. pp. New York: Springer

[9] Kunz Jr JD, Kauzmann W. Hydration of Proteins and Polypeptides. Adv. Protein Chem. 1974;**28**:239–345

[10] Ball P. Water is an active matrix of life for cell and molecular biology. Proc. Natl. Acad. Sci. U.S.A. 2017;**114**: 13327–13335

[11] Levy Y, Onuchic JN. Water Mediation in Protein Folding and Molecular Recognition. Ann. Rev. Biophys. Biomol. Struc. 2006;**35**: 389–415

[12] Chaplin M. Do we underestimate the importance of water in cell biology? Nat. Rev. Mol. Cell Biol. 2006;**7**:861–866

[13] Levitt M, Sharon R. Accurate simulation of protein dynamics in solution. Proc. Natl. Acad.Sci. USA 1988;**85**:7557–7561

[14] Svergun D, Richard S, Koch MH, Sayers Z, Kuprin S, Zaccai G. Protein hydration in solution: Experimental observation by x-ray and neutron scattering. Proc. Natl. Acad. Sci. USA 1998; **95**:2267–2272

[15] Smith JC, Merzel F, Verma CS, Fischer S. Protein Hydration Water: Structure and Thermodynamics. J. Mol. Liq. 2002;**101**:27–33

[16] Uda Y, Zepeda S, Kaneko F, Matsuura Y, Furukawa Y. Adsorption-Induced Conformational Changes of Antifreeze Glycoproteins at the Ice/Water Interface. J. Phys Chem. B 2007; **111**:14355–14361

[17] Yeh Y, Feeney RE. Antifreeze Proteins: Structures and Mechanisms of Function. Chem. Rev. 1996;**96**:601–618

[18] Skalicky JJ, Sukumaran DK, Mills JL, Szyperski T. Toward Structural Biology in Supercooled Water. J. Am. Chem. Soc. 2000;**122**:3230–3231

[19] Rick SW, Stuart SJ, Berne BJ. Dynamical fluctuating Charge Force Fields: Application to Liquid Water. J. Chem. Phys. 1994;**101**:6141

[20] Persson E, Halle B. Cell Water Dynamics on Multiple Time Scales. Proc. Natl. Acad. Sci. USA 2008;**105**: 6266–6271

[21] Ojha L, Wilhelm MB, Murchie SL, Mcewen AS, Wray JJ, et al. Spectral evidence for hydrated salts in recurring

slope lineae on Mars. Nature Geoscience 2015;**8**: 829–832

[22] Zhong D, Pal SK, Zewail AH. Biological water: A critique. Chem. Phys. lett. 2011;**503**:1–11

[23] Stillinger FH. Water Revisited. Science 1980;**209**:451–457

[24] Chong SH, Ham S. Dynamics of Hydration Water Plays a Key Role in Determining the Binding Thermodynamics of Protein Complexes. Sci. Rep. 2017;7:8744

[25] Gavrilov Y, Leuchter JD, Levy Y. On the coupling between the dynamics of protein and water. Physical Chemistry Chemical Physics 2017;**19**:8243–8257

[26] Fogarty AC, Laage D. Water dynamics in protein hydration shells: The molecular origins of the dynamical perturbation. J. Phys. Chem. B 2014;**118**:7715–7729

[27] Wang Z, Bertrand CE, Chiang WS, Fratini E, Baglioni P, et al. Inelastic X-ray scattering studies of the short-time collective vibrational motions in hydrated lysozyme powders and their possible relation to enzymatic function. J. Phys. Chem. B 2013;**117**:1186–1195

[28] Frauenfelder H, Chen G, Berendzen J, Fenimore PW, Jansson H, et al. A unified model of protein dynamics. Proc. Nat. Acad. Sci. USA 2009;**106**:5129–5134

[29] Tarek M, Tobias DJ. The dynamics of protein hydration water: A quantitative comparison of molecular dynamics simulations and neutron-scattering experiments. Biophys. J. 2000;**79**:3244–3257

[30] Wood K, Plazanet M, Gabel F, Kessler B, Oesterhelt D, et al. Dynamics of hydration water in deuterated purple membranes explored by neutron scattering. Eur. Biophys. J. 2008;**37**: 619–626

[31] Otting G, Liepinsh E. Protein Hydration Viewed by High-Resolution NMR Spectroscopy: Implications for Magnetic Resonance Image Contrast. Acc. Chem. Res. 1995;**28**:171–177

[32] Jorge C, Marques BS, Valentine KG, Wand AJ. Characterizing Protein Hydration Dynamics Using Solution NMR Spectroscopy. Methods Enzymol. 2019;**615**:77–101

[33] Camilloni C, Bonetti D, Morrone A, Giri R, Dobson CM, et al. Towards a Structural Biology of the Hydrophobic Effect in Protein Folding. Sci. Rep. 2016; **6**:28285

[34] Moree B, Connell K, Mortensen RB, Liu CT, Benkovic SJ, Salafsky J. Protein Conformational Changes Are Detected and Resolved Site Specifically by Second-Harmonic Generation. Biophy. J. 2015;**109**:806–815

[35] Vanzi F, Sacconi L, Cicchi R, Pavone FS. Protein conformation and molecular order probed by second-harmonic-generation microscopy. J. Biomed. Opt. 2012;**17**:060901

[36] King JT, Kubarych KJ. Site-Specific Coupling of Hydration Water and Protein Flexibility Studied in Solution with Ultrafast 2D-IR Spectroscopy. J. Am. Chem. Soc. 2012;**134**:18705–18712

[37] Ramos S, Horness RE, Collins JA, Haak D, Thielges MC. Site-specific 2D IR spectroscopy: a general approach for the characterization of protein dynamics with high spatial and temporal resolution. Physical Chemistry Chemical Physics 2019;**21**:780–788

[38] Steinbach PJ, Brooks BR. Protein hydration elucidated by molecular dynamics simulation. Proc. Natl. Acad. Sci. USA 1993;**90**:9135–9139

[39] Lagge D, Elsaesser T, Hynes JT. Perspective: Structure and ultrafast

dynamics of biomolecular hydration shells. Struct. Dyn. 2017;**4**:044018

[40] Titantah JT, Karttunen M. Long-time correlations and hydrophobe-modified hydrogen-bonding dynamics in hydrophobic hydration. J. Am. Chem. Soc. 2012;**134**:9362–9368

[41] Mondal S, Mukherjee S, Bagchi B. Dynamical coupling between protein conformational fluctuation and hydration water: Heterogeneous dynamics of biological water. Chem. Phys. lett. 2017;**683**:29–37

[42] Nibali VC, Havenith M. New Insights into the Role of Water in Biological Function: Studying Solvated Biomolecules Using Terahertz Absorption Spectroscopy in Conjunction with Molecular Dynamics Simulations. J. Am. Chem. Soc. 2014;**136**: 12800–12807

[43] Xu Y, Havenith M. Perspective: Watching low-frequency vibrations of water in biomolecular recognition by THz spectroscopy. J. Chem. Phys. 2015; **143**:170901

[44] Haller EE, E. B. 2000. *USA*

[45] Keilmann F, Shastin VN, Till R. Pulse build-up of the germanium far-infrared laser. App. Phys. Lett. 1991;**58**: 2205–2207

[46] Shastin VN. Hot hole inter-sub-band transition p-Ge FIR laser. Opt. Quantum Electornics 1991;**23**:S111–S131

[47] Bründermann E, Chamberlin DR, Haller EE. Thermal effects in widely tunable germanium terahertz lasers. Appl. Phys. Lett. 1998;**73**:2757

[48] Pełka JB, Tyborb KR, Nietubycc R, Wrochna G. Applications of Free Electron Lasers in Biology and Medicine. Acta Physica Polynica A. 2010;**117**:427–432

[49] Novelli F, Pestana LR, Bennett KC, Sebastiani F, Adams EM, et al. Strong anisotropy in liquid water upon librational excitation using terahertz laser fields. J. Phys. Chem. B 2020;**124**: 4989–5001

[50] Bye JW, Meliga S, Ferachou D, Cinque G, Zeitler JA, Falconer RJ. Analysis of the Hydration Water around Bovine Serum Albumin Using Terahertz Coherent Synchrotron Radiation. J. Phys. Chem. A 2014;**118**:83–88

[51] Bergner A, Heugen U, Bründermann E, Schwaab G, Havenith M, et al. New p-Ge THz laser spectrometer for the study of solutions: THz absorption spectroscopy of water. Rev. Sci. Intruments 2005;**76**:063110

[52] Ginsburg V, Frank I. Radiation of uniformly moving electron due to its transition from one medium into another. J. Phys. 1945; **9**:353

[53] Naumenko GA, Aleinik AN, Aryshev AS, Kalinin BN, Potylitsyn AP, et al. Coherent transition and diffraction radiation from a bunched 6.1MeV electron beam. Nuclear Instruments Methods Phys. Res. B 2005;**227**:70–77

[54] Kuroda R, Sei N, Yasumoto M, Toyokawa H, Ogawa H, et al. Generation of 0.1 THz coherent synchrotron radiation with compact S-band linac at AIST. Infrared Phys. Technol. 2008;**51**

[55] Saisut J, Chaisueb N, Thongbai C, Rimjaem S. Coherent transition radiation from femtosecond electron bunches at the accelerator-based THz light source in Thailand. Infrared Physics & Technology. 2018; **92**, 387–391

[56] Ter-Mikaelian ML. *High energy electromagnetic process in condensed media.*1972. pp. New York: Wiley-Intersciencer

[57] Ginzburg VL, Tsytovich VN, eds. 1990. *Transition radiation and transition scattering*. Bristol: Adam Hilger

[58] Kuroda R, Toyokawa H, Sei N, Yasumoto M, Ogawa H, et al. Injector study for compact hard X-Ray source via laser Compton scattering. Int. J. Mod. Phys. 2007;**21**:488

[59] Manna B, Nandi A, Tanaka M, Toyokawa H, Kuroda R, Palit DK. Effect of aggregation on hydration of HSA protein: Steady-state terahertz absorption spectroscopic study. J. Chem. Sci. 2020;**132**:8

[60] Neu J, Schmuttenmaer CA. Tutorial: An introduction to terahertz time domain spectroscopy (THz-TDS). J. App. Phys. 2018;**124**:231101

[61] Roux J-F, Garet F, Coutaz J-L. 2014. Principles and applications of THz time domain spectroscopy. In *Physics and Applications of Terahertz Radiation*, ed. M Perenzoni, DJ Paul. Dordrecht: Springer. Number of

[62] Wang C, Gong J, Xing O, Li Y, Liu F, et al. Application of terahertz time-domain spectroscopy in intracellular metabolite detection. J. Biophotonics 2010;**3**:641–645

[63] Zeitler AJ, Philip TF, David AN, Pepper M, Keith GC, Rades T. Terahertz pulsed spectroscopy and imaging in the pharmaceutical setting–a review. J. Pharm. Pharmacol. 2007;**59**:209–223

[64] Mittleman DM, Jakobsen RH, Neelamani R, Baraniuk RG, Nuss MC. Gas sensing using terahertz time-domain spectroscopy. App. Phys. B 1998;**67**:379

[65] Cheville RA, Grischkowsky D. Far-infrared foreign and self-broadened rotational linewidths of high-temperature water vapor. J. Op. Soc. Am. B 1999;**16**:317

[66] Kužel P, Petzelt J. Time-resolved terahertz transmission spectroscopy of dielectrics. Ferroelectrics 2000;**239**: 949

[67] Jeon T-I, Grischkowsky D. Characterization of optically dense, doped semiconductors by reflection THz time domain spectroscopy. App. Phys. Lett. 1998;**72**:3032

[68] Wilke I, Khazan M, Rieck CT, Kužel P, Kaiser T, et al. Terahertz surface resistance of high temperature superconducting thin films. J. App. Phys. 2000;**87**:2984

[69] Plusquellic DF, Siegrist K, Heilweil EJ, Esenturk O. Applications of Terahertz Spectroscopy in Biosystems. ChemPhysChem 2007;**8**:2412–2431

[70] Schmuttenmaer CA. Exploring Dynamics in the Far-Infrared with Terahertz Spectroscopy. Chem. Rev. 2004;**104**:1759–1780

[71] Beard MC, Turner GM, Schmuttenmaer CA. Terahertz Spectroscopy. J. Phys. Chem. B 2002; **106**:7146–7159

[72] Baxter JB, Guglietta GW. Terahertz Spectroscopy. Anal. Chem. 2011;**83**: 4342–4368

[73] Yu L, Hao L, Meiqiong T, Jiaoqi H, Wei L, et al. The medical application of terahertz technology in non-invasive detection of cells and tissues: opportunities and challenges. RSC Adv. 2019 **9**:9354–9363

[74] Yoon SA, Cha SH, Jun SW, Park SJ, Park J-Y, et al. Identifying different types of microorganisms with terahertz spectroscopy. Biomed Opt Express. 2020;**11**:406–416

[75] Yomogida Y, Sato Y, Nozaki R, Mishina T, Nakahara J. Dielectric study of normal alcohols with THz time-domain spectroscopy. J. Mol. Liq. 2010; **154**:31–35

[76] Sarkar S, Saha D, Banerjee S, Mukherjee A, Mandal P. Broadband terahertz dielectric spectroscopy of

alcohols. Chem. Phys. lett. 2017;**678**: 65–71

[77] Kindt JT, Schmuttenmaer CA. Far-Infrared Dielectric Properties of Polar Liquids Probed by Femtosecond Terahertz Pulse Spectroscopy. J. Phys. Chem. 1996;**100**:10373–10379

[78] Nee TW, Zwanzig R. Theory of dielectric relaxation in polar liquids. The Journal of Chemical Physics 1970;**52**: 6353–6363

[79] Nandi N, Bhattacharyya K, Bagchi B. Dielectric relaxation and solvation dynamics of water in complex chemical and biological systems. Chemical Reviews 2000;**100**:2013–2046

[80] Agieienko V, Horinek D, Buchner R. Hydration and self-aggregation of a neutral cosolute from dielectric relaxation spectroscopy and MD simulations: the case of 1, 3-dimethylurea. Physical Chemistry Chemical Physics 2017;**19**:219–230

[81] Yuri Feldman PBI, Alexander Puzenko, Valerică Raicu. 2015. Elementary Theory of the Interaction of Electromagnetic Fields with Dielectric Materials. Oxford: Oxford University Press

[82] Fröhlich H. *Theory of dielectrics: dielectric constant and dielectric loss* 1958. pp.: Clarendon Press

[83] Bötcher C, Bordewijk P. 1978. Theory of Electric Polarization, vol. II, Dielectrics in Time-Dependent Fields. Elsevier Science

[84] Feldman Y, Puzenko A, Ryabov Y. Dielectic Relaxation Phenomena in Complex Materials. Advances in chemical physics 2006;**133**:1

[85] Schmidt DA, Birer Ö, Funkner S, Born B, Gnanasekaran R, et al. Rattling in the cage: ions as probes of sub-ps water network dynamics. J. Am. Chem. Soc. 2009;**131**:18512–18517

[86] Funkner S, Niehues G, Schmidt DA, Heyden M, Schwaab G, et al. Watching the Low-Frequency Motions in Aqueous Salt Solutions: The Terahertz Vibrational Signatures of Hydrated Ions. J. Am. Chem. Soc. 2012;**134**:1030–1035

[87] Smiechowski M, Sun J, Forbert H, Marx D. Solvation shell resolved THz spectra of simple aqua ions - distinct distance- and frequency-dependent contributions of solvation shells. Physical Chemistry Chemical Physics 2015;**17**:8323–8329

[88] Sharma V, Boehm F, Seitz M, Schwaab G, Havenith M. From solvated ions to ion-pairing: a THz study of lanthanum(III) hydration. Physical Chemistry Chemical Physics 2013;**15**: 8383–8391

[89] Sharma V, Boehm F, Schwaab G, Havenith M. The low frequency motions of solvated Mn(II) and Ni(II) ions and their halide complexes. Physical Chemistry Chemical Physics 2014;**16**: 25101–25110

[90] Boehm F, Sharma V, Schwaab G, Havenith M. The low frequency modes of solvated ions and ion pairs in aqueous electrolyte solutions: iron(II) and iron (III) chloride. Physical Chemistry Chemical Physics 2015;**17**:19582–19591

[91] Klinkhammer C, Boehm F, Sharma V, Schwaab G, Seitz M, Havenith M. Anion dependent ion pairing in concentrated ytterbium halide solutions. J. Chem. Phys. 2018; **148**:222802

[92] Schwaab G, Sebastiani F, Havenith M. Ion Hydration and Ion Pairing as Probed by THz Spectroscopy. Angew. Chem. Int. Ed. 2019;**58**:3000–3013

[93] Das Mahanta D, Samanta N, Mitra RK. The effect of monovalent cations on the collective dynamics of water and on a model protein. J. Mol. Liq. 2016;**215**:197–203

[94] Schienbein P, Schwaab G, Forbert H, Havenith M, Marx D. Correlations in the Solute–Solvent Dynamics Reach Beyond the First Hydration Shell of Ions J. Phys. Chem. Lett. 2017;**8**:2373–2380

[95] Walther M, Fischer BM, Jepsen PU. Noncovalent intermolecular forces in polycrystalline and amorphous saccharides in the far infrared. Chem. Phys. 2003;**261–268**:261–268

[96] Shiraga K, Adachi A, Nakamura M, Tajima T, Ajito K, Ogawa Y. Characterization of the hydrogen-bond network of water around sucrose and trehalose: Microwave and terahertz spectroscopic study. J. Chem. Phys. 2017;**146**:105102

[97] Heugen U, Schwaab G, Bründermann E, Heyden M, Yu X, et al. Solute-induced retardation of water dynamics probed directly by terahertz spectroscopy. Proc. Natl. Acad.Sci.USA 2006;**103**:12301–12306

[98] Lee D-K, Kang J-H, Lee J-S, Kim H-S, Kim C, et al. Highly sensitive and selective sugar detection by terahertz nano-antennas. Sci. Rep. 2015; **5**:15459

[99] Sajadi M, Berndt F, Richter C, Gerecke M, Mahrwald R, Ernsting NP. Observing the Hydration Layer of Trehalose with a Linked Molecular Terahertz Probe. J. Phys. Chem. Lett. 2014;**5**:1845–1849

[100] Heyden M, Bründermann E, Heugen U, Niehues G, Leitner DM, Havenith M. Long range influence of carbohydrates on the solvation dynamics of water - Answers from THz absorption measurements and molecular modeling simulations. J. Am. Chem. Soc. 2008;**130**:5773–5779

[101] Samanta N, Das Mahanta D, Mitra RK. Does Urea Alter the Collective Water Structure: A Dielectric Relaxation Study in THz Region. Chem. Asian J. 2014;**9**:3457–3463

[102] England JL, Haran G. Role of solvation effects in protein denaturation: From thermodynamics to single molecules and back. Ann. Rev. Phys. Chem. 2011;**62**:257–277

[103] Jui-Yoa Chang J-Y. Structural Heterogeneity of 6 M GdmCl-Denatured Proteins: Implications for the Mechanism of Protein Folding. Biochemistry 2009;**48**:9340–9346

[104] Das Mahanta D, Samanta N, Mitra RK. The Decisive Role of Hydrophobicity on the Effect of Alkylammonium Chlorides on Protein Stability: A Terahertz Spectroscopic Finding. J. Phys Chem. B 2017;**121**:7777–7785

[105] Niehues G, Heyden M, Schmidt DA, Havenith M. Exploring hydrophobicity by THz absorption spectroscopy of solvated amino acids. Faraday Discussions 2011;**150**:193–207

[106] Sun J, Niehues G, Forbert H, Decka D, Schwaab G, et al. Understanding THz Spectra of Aqueous Solutions: Glycine in Light and Heavy Water. J. Am. Chem. Soc. 2014;**136**: 5031–5038

[107] Esser A, Forbert H, Sebastiani F, Schwaab G, Havenith M, Marx D. Hydrophilic Solvation Dominates the Terahertz Fingerprint of Amino Acids in Water. J. Phys. Chem. B 2018;**122**:1453–1459

[108] Shiraga K, Suzuki T, Kondo N, Ogawa Y. Hydration and hydrogen bond network of water around hydrophobic surface investigated by terahertz spectroscopy. J. Chem. Phys. 2014;**141**: 235103

[109] Samanta N, Das Mahanta D, Choudhury S, Barman A, Mitra RK. Collective Hydration Dynamics in Some

Amino Acid Solutions: A Combined GHz-THz Spectroscopic Study. J. Chem. Phys. 2017;**146**:125101

[110] Born B, Weingartner H, Bruendermann E, Havenith M. Solvation dynamics of model peptides probed by terahertz spectroscopy. Observation of the onset of collective network motions. J. Am. Chem. Soc. 2009;**131**:3752–3755

[111] Das Mahanta D, Islam SI, Choudhury S, Das DK, Mitra RK, Barman A. Contrasting hydration dynamics in DME and DMSO aqueous solutions: A combined optical pump-probe and GHz-THz dielectric relaxation investigation. J. Mol. Liq. 2019;**290**:111194

[112] Luong TQ, Verma PK, Mitra RK, Havenith M. Onset of Hydrogen Bonded Collective Network of Water in 1,4-Dioxane. J. Phys. Chem. A 2011;**115**: 14462–14469

[113] Das Mahanta D, Patra A, Samanta N, Luong TQ, Mukherjee B, Mitra RK. Non-monotonic dynamics of water in its binary mixture with 1,2-dimethoxy ethane: A combined THz spectroscopic and MD simulation study. J. Chem. Phys. 2016; **145**:164501

[114] Boehm F, Schwaab G, Havenith M. Mapping Hydration Water around Alcohol Chains by THz Calorimetry. Angew. Chem. Int. Ed. 2017;**56**: 9981–9985

[115] Born B, Kim SJ, Ebbinghaus S, Gruebele M, Havenith M. The terahertz dance of water with the proteins: the effect of protein flexibility on the dynamical hydration shell of ubiquitin. Faraday Discussions 2009;**141**:161–173

[116] Heyden M, Havenith M. Combining THz spectroscopy and MD simulations to study protein-hydration coupling. Methods 2010;**52**:74–83

[117] Ebbinghaus S, Kim SJ, Heyden M, Yu X, Gruebele M, et al. Protein Sequence- and pH-Dependent Hydration Probed by Terahertz Spectroscopy. J. Am. Chem. Soc. 2008;**130**:2374–2375

[118] Vondracek H, Dielmann-Gessner J, Lubitz W, Knipp M, Havenith M. THz absorption spectroscopy of solvated β-lactoglobulin. J. Chem. Phys. 2014;**141**: 22D534

[119] Heyden M, Sun J, Funkner S, Mathias G, Forbert H, et al. Dissecting the THz spectrum of liquid water from first principles via correlations in time and space. Proc. Natl. Acad. Sci. USA 2010;**107** 12068–12073

[120] Grossman M, Born B, Heyden M, Tworowski D, Fields GB, et al. Correlated structural kinetics and retarded solvent dynamics at the metalloprotease active site. Nat. Struct. Mol. Biol. 2011;**18**:1102–1108

[121] Adams EM, Lampret O, Koenig B, Happe T, Havenith M. Solvent dynamics play a decisive role in the complex formation of biologically relevant redox proteins. Physical Chemistry Chemical Physics 2020;**22**: 7451–7459

[122] He Y, Ku PI, Knab JR, Chen JY, Markelz AG. Protein Dynamical Transition Does Not Require Protein Structure. Phys. Rev. Lett. 2008;**101**: 178103

[123] He Y, Chen J-Y, Knab JR, Zheng W, Markelz AG. Evidence of Protein Collective Motions on the Picosecond Timescale. Biophys. J. 2011; **100**:1058–1065

[124] Acbas G, Niessen KA, Snell EH, Markelz AG. Optical measurements of long-range protein vibrations. Nat. Commun. 2014;**5**:4076

[125] Turton DA, Senn HM, Harwood T, Lapthorn AJ, Ellis EM, Wynne K.

Terahertz underdamped vibrational motion governs protein-ligand binding in solution. Nat. Commun. 2014;**5**:3999

[126] Niessen KA, Xu M, George DK, Chen MC, Ferré-D'Amaré AR, et al. Protein and RNA dynamical fingerprinting. Nat. Commun. 2019;**10**: 1026

[127] Sun Y, Du P, Lu X, Xie P, Qian Z, et al. Quantitative characterization of bovine serum albumin thin-films using terahertz spectroscopy and machine learning methods. Biomed. Opt. Express 2018;**9**:2917–2929

[128] Sun Y, Junlan Zhong J, Zhang C, Zuo J, Pickwell-MacPherson E. Label-free detection and characterization of the binding of hemagglutinin protein and broadly neutralizing monoclonal antibodies using terahertz spectroscopy. J. Biomed. Optic 2015;**20**:037006

[129] Sun Y, Zhang Y, Pickwell-MacPherson E. Investigating Antibody Interactions with a Polar Liquid Using Terahertz Pulsed Spectroscopy. Biophys. J. 2011;**100**:225–231

[130] Sun Y, Zhu Z, Chen S, Balakrishnan J, Abbott D, et al. Observing the Temperature Dependent Transition of the GP2 Peptide Using Terahertz Spectroscopy. PloS one 2012;**7**: e50306

[131] Markelz AG, Roitberg A, Heilweil EJ. Pulsed terahertz spectroscopy of DNA, bovine serum albumin and collagen between 0.1 and 2.0 THz. Chem. Phys. Lett. 2000;**320**: 42–48

[132] Fischer BM, Walther M, Jepsen PU. Far-infrared vibrational modes of DNA components studied by terahertz time-domain spectroscopy. Phys. Med. Biol. 2002;**47**:3807–3814

[133] Brucherseifer M, Nagel M, Bolivar PH, Kurz H. Label-free probing of the binding state of DNA by time-domain terahertz sensing. Appl. Phys. Lett. 2000;**77**:4049–4051

[134] Arora A, Luong TQ, Kruger M, Kim YJ, Nam C-H, et al. Terahertz-time domain spectroscopy for the detection of PCR amplified DNA in aqueous solution. Analyst 2012;**137**:575

[135] Tang M, Zhang M, Yan S, Xia L, Yang Z, et al. Detection of DNA oligonucleotides with base mutations by terahertz spectroscopy and microstructures. PLoS One 2018;**13**: e0191515

[136] Polley D, Patra A, Mitra RK. Dielectric relaxation of the extended hydration sheathe of DNA in the THz frequency region. Chem. Phys. Lett. 2013 **586**:143–147

[137] Tielrooij KJ, Paparo D, Piatkowski L, Bakker HJ, Bonn M. Dielectric Relaxation Dynamics of Water in Model Membranes Probed by Terahertz Spectroscopy. Biophys. J. 2009;**97**:2484–2492

[138] Yamamoto N, Andachi T, Tamura A, Tominaga K. Temperature and Hydration Dependence of Low-Frequency Spectra of Lipid Bilayers Studied by Terahertz Time-Domain Spectroscopy. J. Phys. Chem. B 2015; **119**:9359–9368

[139] Yamamoto N, Ito S, Nakanishi M, Chatani E, Inoue K, et al. Effect of Temperature and Hydration Level on Purple Membrane Dynamics Studied Using Broadband Dielectric Spectroscopy from Sub-GHz to THz Regions. J. Phys. Chem. B 2018;**122**:1367–1377

[140] Kadomura Y, Yamamoto N, Tominaga K. Broadband dielectric spectroscopy from sub GHz to THz of hydrated lipid bilayer of DMPC. Eur. Phys. J. E 2019 **42**:139

[141] Pal S, Samanta N, Das Mahanta D, Mitra RK, Chattopadhyay A. Effect of

Phospholipid Headgroup Charge on the Structure and Dynamics of Water at the Membrane Interface: A Terahertz Spectroscopic Study. J. Phys. Chem. B 2018;**122**:5066–5074

[142] Kim SJ, Born B, Havenith M, Gruebele M. Real-time detection of protein-water dynamics upon protein folding by terahertz absorption spectroscopy. Angew. Chem. Int. Ed. 2008;**47**:6486–6489

[143] Luong TQ, Verma PK, Mitra RK, Havenith M. Do Hydration Dynamics Follow the Structural Perturbation during Thermal Denaturation of a Protein: A Terahertz Absorption Study Biophys. J. 2011;**101**:925–933

[144] Luong TQ, Xu Y, Bründermann E, Leitner DM, Havenith M. Hydrophobic collapse induces changes in the collective protein and hydration low frequency modes. Chem. Phys. Lett. 2016;**651**:1–7

[145] Wirtz H, Schäfer S, Hoberg C, Reid KM, Leitner DM, Havenith M. Hydrophobic Collapse of Ubiquitin Generates Rapid Protein–Water Motions. Biochemistry 2018;**57**:3650–3657

[146] Li X, Fu X, Liu J, Du Y, Hong Z. Investigation of thermal denaturation of solid bovine serum albumin by terahertz dielectric spectroscopy. J. Mol. Struct. 2013;**1049**:441–445

[147] Zang Z, Yan S, Han X, Wei D, Cui H-L, Du C. Temperature- and pH-dependent protein conformational changes investigated by terahertz dielectric spectroscopy. Infrared Phys. Technol. 2019;**98**:260–265

[148] Cao C, Serita K, Kitagishi K, Murakami H, Zhang Z-H, Tonouchi M. Terahertz Spectroscopy Tracks Proteolysis by a Joint Analysis of Absorptance and Debye Model. Biophys. J. (DOI: 10.1016/j.bpj.2020.11.003) 2020

[149] Samanta N, Luong TQ, Das Mahanta D, Mitra RK, Havenith M. Effect of Short Chain Poly(ethylene glycol)s on the Hydration Structure and Dynamics around Human Serum Albumin. Langmuir 2016;**32**:831–837

[150] Das DK, Patra A, Mitra RK. Preferential solvation of lysozyme in dimethyl sulfoxide/water binary mixture probed by terahertz spectroscopy. Biophys. Chem. 2016;**216**: 31–36

[151] Das DK, Das Mahanta D, Mitra RK. Non-monotonous Hydration Behaviour of Bovine Serum Albumin in Alcohol-Water Binary Mixtures: A THz Spectroscopic Investigation. ChemPhysChem 2017;**18**:749–754

[152] Stefani M, Dobson CM. Protein aggregation and aggregate toxicity: new insights into protein folding, misfolding diseases and biological evolution. J. Mol. Med. 2003;**81**:678–699

[153] Ross CA, Poirier MA. Protein aggregation and neurodegenerative disease. Nat. Med. 2004;**10**:S10-S17

[154] Zhang SG. Fabrication of Novel Biomaterials through Molecular Self-Assembly. Nat. Biotechnol. 2003;**21**: 1171–1178

[155] Gotz J, Ittner LM, Lim YA. Common features between diabetes mellitus and Alzheimer's disease. Cellular Mol. Lifesci. 2009;**66**:1321–1325

[156] J.C. R, Outeiro TF, Conway KA, Ding TT, Volles MJ, et al. Interactions among alpha-synuclein, dopamine, and biomembranes-some clues for understanding neurodegeneration in Parkinson's disease. J. Mol. Neurosci. 2004;**23**:23–33

[157] Sa'a P, Harris DA, Cervenakova L. Mechanisms of prion-induced neurodegeneration. Expt. Rev. Mol. Med. 2016;**18**:e5

[158] Nayak A, Dutta AK, Belfort G. Surface-enhanced nucleation of insulin amyloid fibrillation. Biochem. Biophysics Res. Comm. 2008;**369**:303–307

[159] Chiti F, Dobson CM. Protein misfolding, functional amyloid, and human disease. Annu. Rev. Biochem. 2006;**75**:333–366

[160] Fink AL. Protein aggregation: folding aggregates, inclusion bodies and amyloid. Folding & Design 1998;**3**:R9-R23

[161] Poulson BG, Szczepski K, Lachowicz JI, Jaremko L, Emwas AH, Jaremko M. Aggregation of biologically important peptides and proteins: inhibition or acceleration depending on protein and metal ion concentrations. RSC Adv. 2020;**10**:215–227

[162] Chong S-H, Ham S. Interaction with the Surrounding Water Plays a Key Role in Determining the Aggregation Propensity of Proteins. Angew. Chem. Int. Ed. 2014;**53**:3961–3964

[163] Chong SH, Ham S. Impact of chemical heterogeneity on protein self-assembly in water. Proc. Natl. Acad. Sci. USA 2012;**109**:7636–7641

[164] Kodaka M. Interpretation of concentration-dependence in aggregation kinetics. Biophys. Chem 2004;**109**:325–332

[165] Bhattacharya A, Prajapati R, Chatterjee S, Mukherjee TK. ConcentrationDependent Reversible Self-Oligomerization of Serum Albumins through Intermolecular β-Sheet Formation. Langmuir 2014;**30**: 14894–14904

[166] Huang K, Kingsley CN, Sheil R, Cheng C, Bierma JB, et al. Stability of protein-specific hydration shell on crowding. J. Am. Chem. Soc. 2016;**138**: 5392–5402

[167] Molodenskiy D, Shirshin E, Tikhonova T, Gruzinov A, Petersa G, Spinozzi F. Thermally induced conformational changes and protein-protein interactions of bovine serum albumin in aqueous solution under different pH and ionic strengths as revealed by SAXS measurements. Physical Chemistry Chemical Physics 2017;**19**:17143–17155

[168] Baler K, Martin OA, Carignano MA, Ameer GA, Vila JA, Szleifer I. Electrostatic Unfolding and Interactions of Albumin Driven by pH Changes: A Molecular Dynamics Study. J. Physs. Chem. B 2014;**118**:921–930

[169] Ebbinghaus S, Kim SJ, Heyden M, Yu X, Heugen U, et al. An extended dynamical hydration shell around proteins. Proc. Natl. Acad. Sci. USA 2007;**104**:20749–20752

[170] Novelli F, Ostovar Pour SO, Tollerud J, Roozbeh A, Appadoo DRT, et al. Time-Domain THz Spectroscopy Reveals Coupled Protein–Hydration Dielectric Response in Solutions of Native and Fibrils of Human Lysozyme. J. Phys. Chem. B 2017;**121**:4810–4816

[171] Chong S-H, Ham S. Distinct Role of Hydration Water in Protein Misfolding and Aggregation Revealed by Fluctuating Thermodynamics Analysis. Acc. Chem. Res. 2015, 4;**48**:956–965

[172] Dorbez-Sridi R, Cortès R, Mayer E, Pin S. X-ray scattering study of the structure of water around myoglobin for several levels of hydration. J. Chem. Phys. 2002;**116**:7269

[173] Sushko O, Dubrovka R, Donnana RS. Sub-terahertz spectroscopy reveals that proteins influence the properties of water at greater distances than previously detected. J. Chem. Phys. 2015;**142**:055101

[174] Patro SY, Przybycien TM. Simulations of reversible protein aggregate and crystal structure. Biophys. J. 1996;**70**:2888–2902

[175] Jachimska B, Wasilewska M, Adamczyk Z. Characterization of Globular Protein Solutions by Dynamic Light Scattering, Electrophoretic Mobility, and Viscosity Measurements. Langmuir 2008;**24**:6866–6872

Chapter 5

THz Imaging for Food Inspections: A Technology Review and Future Trends

Sonia Zappia, Lorenzo Crocco and Ilaria Catapano

Abstract

Terahertz imaging is the newest among non-invasive sensing technologies and currently huge attention is pointed towards its use in several applications. Among possible applications, food inspection represents one of the most prominent cases, due to the possible dangerous impact on human safety. Hence, significant efforts are currently addressed towards the exploitation of THz imaging as a tool to improve the effectiveness of food quality surveys. This chapter deals with the exploitation of THz imaging technology for food quality control and assessment. In particular, the chapter aims at reviewing the latest developments regarding THz imaging, both in terms of measurement systems and data processing methodologies. Moreover, the chapter summarizes experiments available in literature and presents some purposely designed experiments to address the discussion on currently open issues.

Keywords: Terahertz waves, Teraherz imaging, Food quality, Foreign substance, Non Destructive evaluation

1. Introduction

The agricultural and food (agri-food) industry has always attracted significant attention worldwide. In particular, the importance of reliable diagnostic inspection technologies has been growing due to an increasing demand for improving the quality of life. Contamination by foreign bodies, packaging failures and the production of articles with poor characteristics (consistency, appearance) are among the main sources of customer complaints against manufacturing companies, with consequent loss of brand credibility. The food industry is particularly exposed to this problem, especially nowadays that consumers are more aware and pay much more attention to the quality and integrity of the food purchased. Accordingly, huge interest is towards the use of sensing technologies capable of detecting foreign body contamination and packaging failures. Available sensing techniques for food inspection include X-ray imaging [1–3], thermal imaging [4, 5], ultrasonic imaging [6, 7], fluorescence imaging [8] and electromagnetic (EM) systems working at microwave frequencies [9–12]. However, each one of these techniques has advantages and limitations. For example, X-rays systems, which are increasingly used in the food production industry for quality control inspection, provide high resolution images but have difficulties in detecting low density objects such as plastic, glass, wood or insects. Furthermore, the use of ionizing radiation is always related to risks involving both operators and the food itself that could be altered. On the other hand, infrared (IR) technologies have the advantage of being fast and safe but they are limited by a poor penetration capability and strong absorption in water. Instead, fluorescence imaging is effective only when objects with fluorescent compounds are investigated. In this framework, terahertz (THz) imaging appears as an emerging technology, which offers several advantages. THz are electromagnetic waves ranging from 0.1 to 30 THz (wavelength from 3 *mm* to 10 μ*m*), which are currently exploited in various applications, including medical diagnosis [13], pharmaceutical analysis [14], security enhancement [15] and artwork [16]. As microwaves and infrared signals, THz waves are non ionizing radiations thus they allow a safe survey without requiring specific security protocols. Moreover, similar to microwave imaging and differently from X-ray technology, THz systems allow the detection and localization of changes in terms of electromagnetic properties with a high resolution (generally speacking, THz spatial resolution is in the order of some hundred microns while the microwave spatial resolution is typically about centimiters). On the other hand, THz capability of penetrating inside a material is limited to a few millimiters. In addition, THz waves are strongly attenuated by water and suffer the environment humidity conditions. Currently, several studies regarding the detection of both food contaminants [17, 18] and defects in plastic packages [19] have shown the effectiveness of THz imaging technology in the field of food industry. However, the study of THz potential in food inspection is still at an early stage and far to be completely assessed.

This chapter provides an overview of THz imaging for food inspection by summarizing the current state of the technological development and recalling valuable case studies presented in literature. For sake of completeness, the chapter starts with a brief description of the main THz technologies, i.e. THz pulsed and time domain systems as well as continuous wave ones. Then, THz imaging in transmission mode, THz Time of Flight imaging and THz Camera imaging are reviewed and their advantages and drawbacks are discussed. Thereafter, the attention is focused on the employment of THz imaging for food quality control and its potentialities in detecting contaminants and packaging failures. Relevant examples available in literature are described together with some experiments carried out by the authors themselves. Finally, the main open challenges and future perspectives are discussed.

2. THz imaging systems

Systems for THz imaging can be divided into two main categories based on their operative principle: pulsed and continuous wave systems.

Pulsed or time domain (TD) systems are composed of four principal components: primary source, which is an ultrafast pulsed laser emitting sub 100 *fs* pulses, THz emitter, THz detector and a time delay stage.

There are two major technologies to generate and detect THz pulses: photoconductive antennas (PCAs) and electro-optic crystal (EOC); usually one emitter and one detector are used, even if technologies based on the use of detectors arrays are available [20]. The ultrafast laser beam is divided into pump and probe beams (**Figure 1**). THz pulse is generated by the pump beam that affects the emitter, while the probe beam is used to gate the detector. The probing THz wave is focused onto the sample using polymeric lenses or mirrors. After the interaction with the sample, the THz wave is collimated and refocused on the terahertz detector. The delay stage is used to offset the pump and probe beams and allows the temporal sampling of the THz signal.

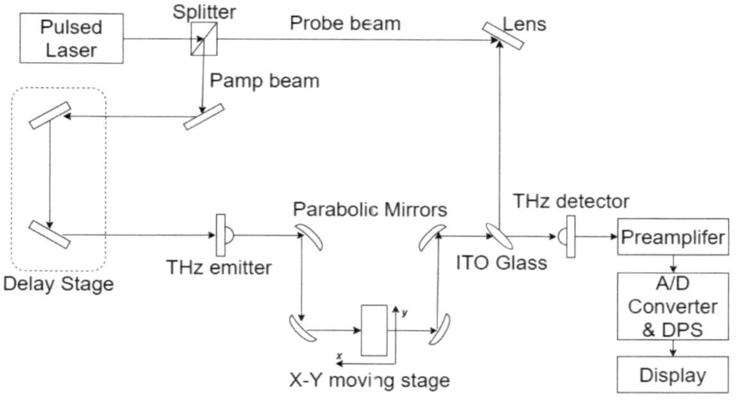

Figure 1.
Typical transmission THz imaging system.

CW or frequency domain systems are based on the beating of two laser wavelengths, whose difference is in the THz region, in a biased semiconductor photomixer [21]. If the photomixer is connected to a properly designed antenna, THz radiation is emitted into the free space. The CW THz radiation is then collimated and guided in the same manner as for pulsed systems [21]. TD and CW THz systems have many differences. TD systems are more complex, expensive and heavy than CW ones but they work in a wide band, usually from 0.1 up to 6 THz for the new generation systems [22]. Moreover, they gather the signal as a function of the time and, thus, allow to reconstruct more information about the investigated samples, provided sophisticated data processing procedures are adopted. Conversely, CW systems are narrow-band and often gather only the intensity of the detected signals, which is stored in a matrix and directly converted to a raster image. Therefore, they are effective in generating THz images almost in real-time but also make inaccessible some information and thus are less flexible. It is worth remarking that TD systems encode depth information as changes in pulse timing, i.e. a variation of the temporal location of the measured peak that indicates a change in the optical path length from the emitter to the sample and from this latter to the detector. CW systems also make possible to obtain this information through the use of additional detectors or by performing additional scans. For instance, by changing the angle between emitter and detector, it is possible to measure the scattered wave instead of the specular reflection. Hence, by performing measurements at different angles it is possible to record intensity data accounting for scattering and reflection

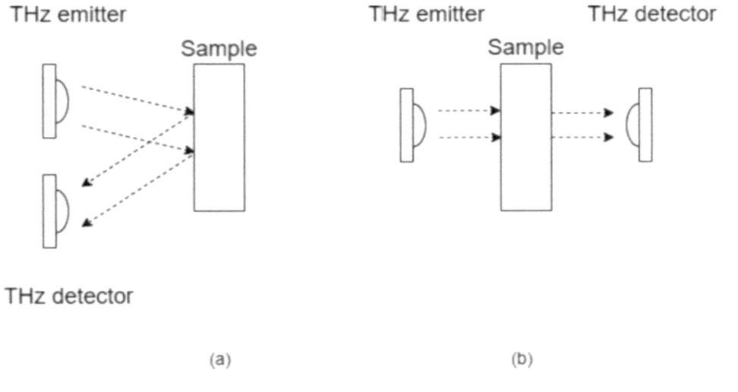

Figure 2.
THz setup. (a) Reflection mode. (b) Transmission mode.

phenomenon occurring at different depths and, thus, to image features of each layer of the measured sample [23].

The choice between TD or CW systems depends on the application of interest. Focusing on food inspections, a preferable choice is a TD system, when the goal is the non-destructive inspection devoted to detect defects or foreign bodies, because it provides more complete information and make possible a 3D visualization of the investigated sample. However, a CW system is sufficient when a 2D image is enough to satisfy the survey finality, as it happens, e.g., when the control of packaging is considered.

Finally, it is worth recalling that both TD and CW systems can be used in reflection or transmission mode as sketched in **Figure 2a** and **b**, respectively.

3. THz imaging modalities

3.1 THz imaging in transmission mode

When working in transmission mode, the THz radiation propagates through the material to be inspected from the emitter to the detector. Hence, in this type of setup one places an object at the focus of the THz beam on an x-y moving stage, and measures the waveform passing through the material [24]. By translating the object and measuring the transmitted THz waveform at each position of the object, a pixel by pixel image is built. The transmitted signal is collected as a time-dependent function that contains information regarding the phase and amplitude of the THz field. The recorded signal is digitized and processed by the digital signal processor (DSP) at each pixel to obtain the transmission THz image of a sample [25]. Specifically, several 2D THz images are obtained by plotting, point by point, different features of the measured wave-forms, i.e. their amplitude, phase, maximum or minimum values and so on [26]. Moreover, by means of the Fast Fourier Transform (FFT), frequency-domain information on the samples are also obtained [27]. The transmission mode is effective because many materials strongly absorb THz radiation, while others are transparent. Hence, it is possible to obtain information on the sample by plotting the amplitude of the measured wave-forms and observing its spatial distribution, while taking into account that higher amplitude values correspond to THz transparent materials. However, if the sample is too thick no useful image can be obtained. Further information is obtained by computing the absorption coefficient and refractive index by using the sample $E_s(\nu)$ and the reference $E_r(\nu)$ spectra. In transmission mode, $E_r(\nu)$ is calculated by taking measurements without the sample. Specifically, the transmission function is defined as follows [28, 29]:

$$T(\nu) = \frac{E_s(\nu)}{E_r(\nu)} = \frac{4n}{(n+1)^2} exp\left[-\alpha\frac{d}{2} + j2\pi\nu(n-1)\frac{d}{c}\right] = Aexp[j\phi(\nu)] \tag{1}$$

where d is the sample thickness, ν is the radiation frequency, c is the speed of light in vacuum, n is the refractive index, α is the absorption coefficient, A is the amplitude ratio between the sample and reference spectra and ϕ is the relative phase difference. Hence, it is possible to calculate n and α for each frequency as:

$$n = 1 + \frac{c\phi}{2\pi\nu d} \tag{2}$$

$$\alpha = \frac{2}{d} ln\left[\frac{4nA}{(n+1)^2}\right] \tag{3}$$

By taking into account that some materials exhibit strong absorption peaks at specific THz frequencies, once the absorption coefficient and the refractive index have been computed, it is possible to collect information on the structure of the sample and on its materials. THz transmission imaging provides a wealth of information by analyzing the 2D images of these indices at different frequencies, but the pixel by pixel scanning procedure is time consuming and often suffers from poor spatial resolution because the spatial offset used to move mechanically the sample is often comparable or larger than the wavelength of the THz radiation [30].

3.2 THz time of flight imaging (TOF)

THz time-of-flight imaging (TOF) also known as THz pulsed imaging (TPI) has the unique property of providing a 3D "map" of the object by exploiting data collected in reflection mode [31]. This technique was used for the first time by Mittleman and coworkers to produce the internal structure of a 3.5 inch floppy disk [24]. The apparatus used for reflection imaging is quite similar to the transmission one, except for the change regarding the THz beam path (see **Figure 2**) [32]. In brief, the object is probed by a pulse signal and the reflected waveform is collected as a time-dependent function in a certain observation time window using a THz-TDS configuration. The temporal delay of the reflected pulses reveals the internal structure of the sample (if the object is nonmetallic). By collecting data along a line, moving emitter and receiver along a straight trajectory and plotting the gathered wave-forms, a space–time image is obtained. In particular, the data can be represented in the form of a THz *radargram* (see **Figures 7** and **8** in Section 4.1), that is a two-dimensional image, wherein the horizontal axis is the spatial coordinate and represents the measurement line, while the vertical axis is the temporal coordinate and represents the time of flight, that is, the time T that the waveform employs to propagate from the emitter to an electromagnetic discontinuity and to go back to the receiver (see **Figure 3**). The time of flight T is related to the distance d between THz probes and the detected discontinuities as:

$$T = \frac{2d}{v} \tag{4}$$

v being the electromagnetic wave propagation velocity into the object. Therefore, a THz radargram provides a cross-sectional representation of the

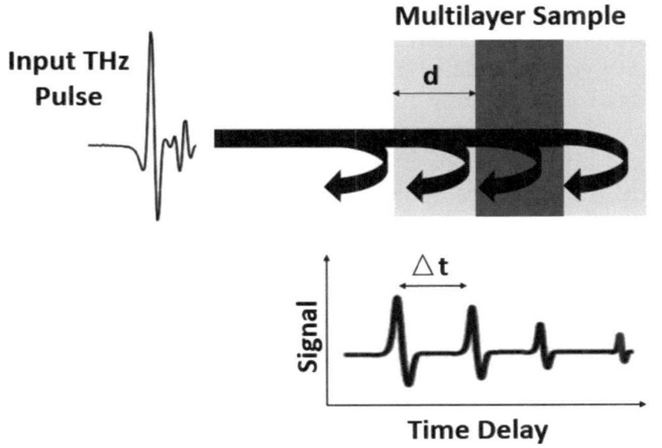

Figure 3.
Time of flight sensing principle.

inner features of the object, which gives information about position and thickness of possible inner discontinuities. A 3D characterization is obtained by collecting radargrams along parallel profiles and visualizing the stored data all together.

Reflection mode is effective because different materials exhibit different behavior when illuminated by THz radiation. In principle, as in transmission mode, the spatial distribution of the refractive index can be retrieved, since amplitude and phase of the reflected signals are available by means of the Fourier transform, even if it is quite complex to define the reference signal and this makes the calculation of the reflectivity more complex. On the other hand, it is important to take into account that the reflectivity of an object depends not only on its refractive index, but also on the surface roughness, polarization of incident THz radiation and other geometric factors [15]. In addition, noise is a crucial problem in this case and advanced filtering and signal processing approaches are required. In fact, the intensity of the signal measured in reflection mode is generally lower than in transmission mode. In this respect, noise filtering and deconvolution procedures are commonly applied to improve the final image quality [33].

3.3 THz camera imaging system

An important research topic regards the development of THz cameras similar to the optical ones. Although active and passive THz cameras have been proposed, THz cameras are mainly designed as an active sensor, being the natural emission of THz wavelength extremely weak. Therefore, an appropriate source has to be used to illuminate the scene. Two main types of THz cameras are available in literature: thermal cameras and FET cameras [34]; they are briefly described in the following.

Thermal cameras record an image based on the heat generated by the THz radiation and exploit three types of thermal detectors: Golay cell, Pyroelectric sensors and bolometers. Golay cells transfer the generated heat to an expanding gas cell and measure the increased pressure optically by means of moving mirrors [35]. Being based on the mechanical movement of a membrane, Golay cells are slow detectors and are difficult to be integrated into dense arrays of detectors. Consequently, THz imaging with Golay cells is often done in a single-pixel detection scheme. Pyroelectric sensors detect changes in terms of the electrical polarizability of certain crystals, which is generated by a temperature growth [36]. Unlike Golay cells, pyroelectric cameras based on an array of detectors are commercially available but are characterized by a low sensitivity and this issue makes them less effective for THz imaging. Bolometers detect the change in a temperature-dependent resistance and are often based on the well-known infrared imaging technology adapted to THz frequencies [37]. Currently, microbolometers (i.e arrays of bolometers mounted onto read-out integrated circuits fabricated with CMOS process technology) are specifically designed to operate at THz frequencies and make possible to reach a sensitivity level suitable for industrial imaging applications, thanks to the inclusion of a Fabry–Perot cavity, antennas, and metamaterials to increase the absorbed THz radiation [38].

THz FET-based cameras measure a rectified voltage after interaction of the THz radiation with a plasma wave in a transistor channel [39]. This approach based on antenna-coupled FETs for THz detection have been implemented in standard CMOS technology [40]. Their production does not require different procedures with respect to standard CMOS technology, which provides the benefit of high maturity in terms of fabrication yield and pixel uniformity. On the other hand, to perform a successful THz imaging, the source beams must be expanded and in several standard setup [41] such an expansion is realized by means of parabolic mirrors without any absorption.

Generally specking, the operating principle of a THz camera does not differ so much from that described in the previous paragraphs: a source illuminates the object under test and the reflected (or transmitted) signal is collected by the camera. Moreover, between the object under test and the camera there are often lenses devoted to focus the THz beam on the camera. The main difference with the THz imaging system described in 3.1 and 3.2 is that the THz cameras provide non-discriminatory intensity-based imaging. It is worth pointing out that, at the current state of the technological development, THz cameras appear suitable to a possible industrial application even if there are still issues related to the requirements to be satisfied. First, the cameras manufacturing process should be compatible with existing manufacturing techniques, such as CMOS technology, which enables the development of high resolution image arrays, in order to reduce the costs. THz cameras must be able to operate at room temperature while still retaining high sensitivity and should be characterized by contained size, weight and energy consumption in order to facilitate integration into industrial imaging systems. Furthermore, it is crucial to assure that data acquisition and processing time are compliant with the time of a production line [42, 43].

4. Applications of THz imaging in food industry

This paragraph provides an overview of the potentialities offered by THz imaging in the frame of food inspections by presenting examples available in literature as well as experiments carried out by the authors themselves. Specifically, two kind of inspections are considered: 1) the detection of defect and foreign bodies; 2) the packaging quality control.

It is worth pointing out that non-polar and non-metallic substances, such as plastic, cardboard, wood and other common packaging materials, are transparent to THz radiation and show very weak interaction with THz waves. This makes THz imaging attractive tools for the inspections of packaged products. Consequently, the use of THz waves allows us to detect unwanted objects that compromise the safety of the product as well as to check defects of the sample under test by controlling its integrity. In this way, defects such as irregular shape, breaking and even absence of the product itself inside the package can be detected.

4.1 Defect and foreign body detection

The presence of defects and foreign bodies in food is one of the major concerns of the food industry. Usually, the term 'foreign body' refers to unwanted objects in food products and it accounts for contaminants originated both by food products themselves (e.g bone in meat; fruit stones) and by manufacturing or packaging processes (e.g metal, plastic, glass pieces and small stones) as well as for warms and insects. In the last years, several research groups have focused their activities on THz surveys devoted to detect and localize undesired hidden objects and to verify the quality of food products. For this purpose, both CW systems and pulsed ones have been employed.

The employment of THz systems for the detection of contaminants such as small stones, glass, plastic and metal has been considered to investigate flour [44], milk powder [45], wheat grain [46] and chocolate bars [47]. It is worth reporting, as an example, the THz inspection of chocolate bars [17], that represents a promising field of THz imaging application. In [17], THz imaging was performed in transmission mode with the aim of detecting contaminants such as stones, glass and metal screws in chocolate bars and a THz-TDS working in the frequency range from 50 GHz to 2.5 THz was used. The 2D images were obtained by scanning the samples perpendicularly to the THz beam. In all the analyzed cases, contaminants were clearly localized in THz images due to the lower intensity of the transmitted THz radiation when the foreign body were present. In fact, glass and stone absorbed part of the signal decreasing its intensity, while metal reflected all THz radiation. THz transmission imaging allowed to identify foreign bodies in the chocolate sample both in the presence and absence of the plastic foil packaging. In [17], it has been also demonstrated that is sufficient to record only some key points of the THz waveform in order to collect all the necessary information thus decreasing the amount of data and the measurement time and bringing the scanning speeds to 0.55 m/s. Furthermore, THz technology has been exploited to detect undesired hidden insects such us crickets, mealworms and grasshoppers in flour [48], milk powdered [18], noodle [49–51] and chocolate bars [47]. These studies have been referred to laboratory prepared contaminated samples and have provided a preliminary assessment of THz capabilities to detect insects in food.

The possibility to check the quality of red ginseng and walnuts has been investigated in [52], where a sub wavelength transmission imaging system based on Bessel–Gauss beam focusing in the 140 GHz frequency band has been used. Information about the quality of the analyzed product was obtained since THz images showed that high quality samples have a relatively uniform internal structure. It has been demonstrated that red ginseng and walnuts with thicknesses of 20–30 mm could be inspected without causing damage and the internal structure of these objects was observed non destructively. Finally, already commercially available THz imaging system [53], has been used to check the presence (or the absence) of chocolate bars inside their package, preventing any complaints due to an incorrect number of bars.

In this context, the potentialities offered by THz imaging has been assessed by the authors themselves and two test cases involving chocolate cream are herein presented. The results have been obtained by means of the Zomega THz FiCo system available at the Institute for Electromagnetic Sensing of the Environment - National Research Council of Italy (IREA-CNR). The adopted THz system is equipped with an ad hoc designed imaging module, which allows an automatic planar scan (see **Figure 4**) and collects data in normal reflection mode in the effective frequency range from 40 GHz to 1.16 THz. For a more precise description of the system see [16, 54]. The collected data have been processed by means of a two-step filtering procedure: a band pass filter designed to select the effective spectrum of the signal [54] followed by a singular value decomposition (SVD) procedure [16].

Figure 5 shows the chocolate laboratory-prepared samples that are made by a plastic support filled with a chocolate cream: in the first case there is a surface defect consisting of a piece of pistachio peel (see 5a), in the second case metal has hidden 1 mm deep inside the chocolate cream, mimicking the foreign body (see 5b). False-color THz images are shown in **Figure 6**, and have been obtained by plotting, point by point, the maximum amplitude of the processed data collected by the system. **Figure 6** demostrates the THz imaging ability to detect pistachio peel on the surface of the chocolate cream (see 6a) and the presence of the metal inside the sample (see 6b). **Figure** 7 shows filtered data, referred to the sample in **Figure 5a**, that have been represented in the form of a THz radargram as described in Sec 3.2. The data have been gathered with a 0.25 mm spatial step along the x-axis, for y = 14 mm and y = 20 mm, which correspond to lines where the surface defect is present and absent, respectively. By observing 7a, it is possible to note that the extent of the

Figure 4.
Zomega FiCO system: Imaging module.

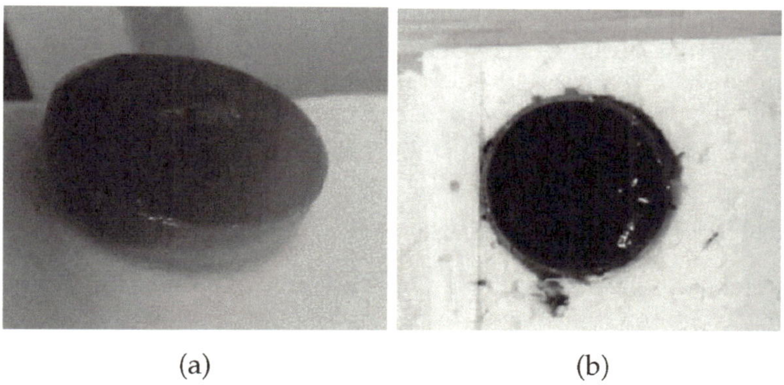

Figure 5.
Laboratory-prepared samples. (a) Surface defect. (b) Foreign body covered by a layer of chocolate.

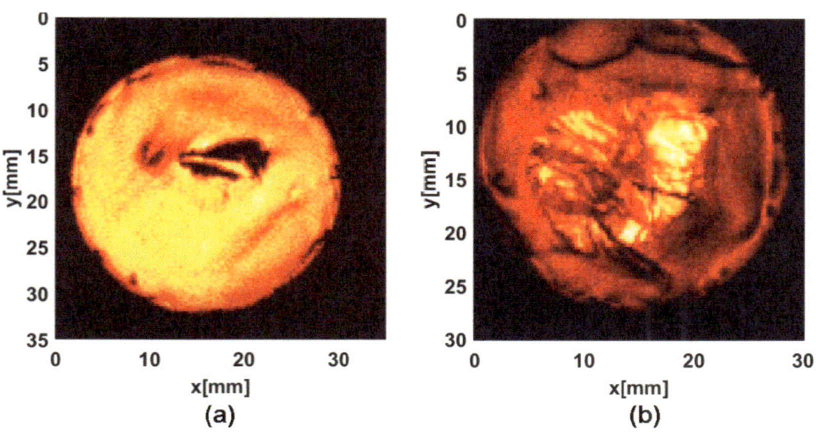

Figure 6.
False color THz images. (a) Surface defect detection. (b) Foreign body detection.

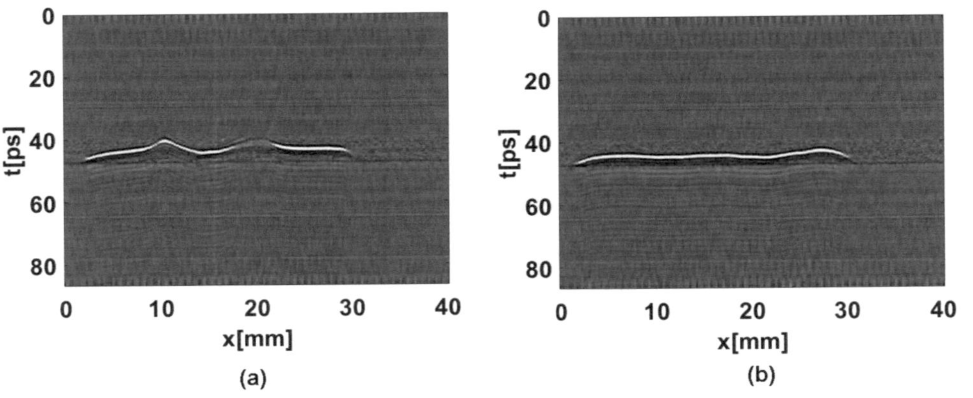

Figure 7.
THz Radargrams referred to 5a. (a) y = 14 mm - the extent of the surface defect ranges from 10 mm to 22 mm. (b) y = 20 mm - No defect.

defect ranges from 10 mm to 22 mm. **Figure 8** shows THz radargram, referred to the sample in **Figure 5b**, gathered with a 0.25 mm spatial step along the x-axis, for $y = 20$ mm, that corresponds to an area where the contaminant is present. By observing **Figure 8**, it is possible to distinguish two reflections due to occurrence of different materials. In particular, the first reflection is related to the air-chocolate interface, while the second reflection represents the interface between the chocolate and the metal layer. Based on the obtained results, one can state that THz waves allow us not only to image the surface defect on the food samples, but also to detect and localize hidden objects that may be dangerous for human life.

4.2 Packaging quality control

The detection of production defects in plastic or paper packaging is a potential area of application for THz imaging. The possibility to check the integrity of plastic welded joints using THz waves has been explored in [55] where a THz - TDS system in transmission mode with a frequency range from 50 GHz to 2.5 THz has been used. Several sheets of high-density polyethylene welded into a lap joint with

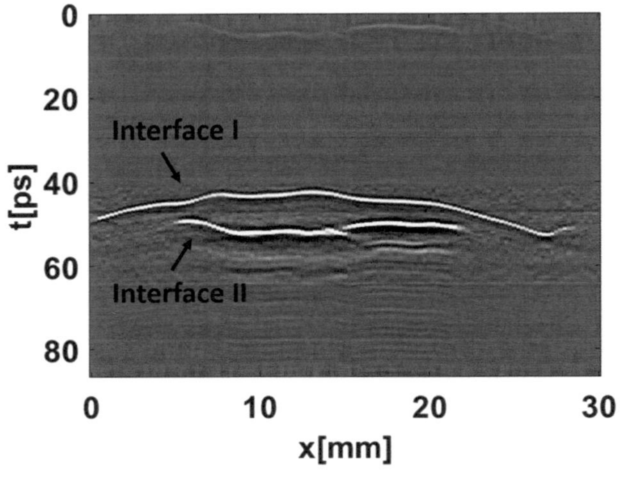

Figure 8.
THz Radargram referred to 5b - y = 11 mm. It is possible to distinguish two reflections due to occurrence of different materials.

Figure 9.
Sugar bags. (a) Standard packaging. (b) Defect packaging highlighted in the box.

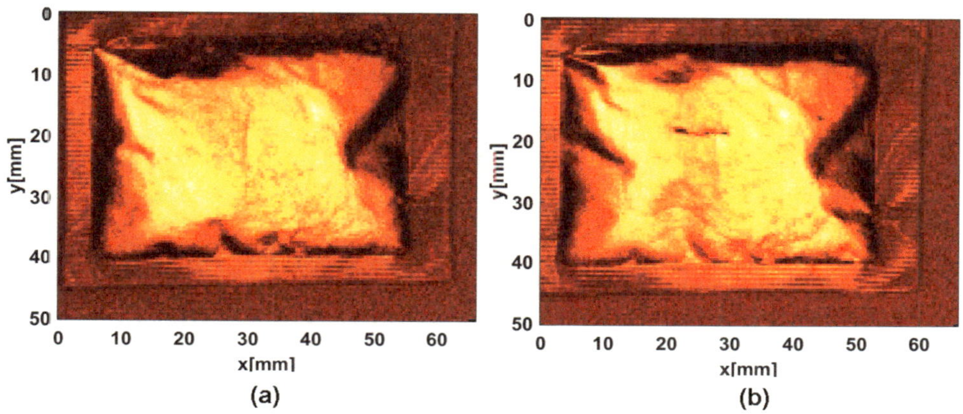

Figure 10.
False color THz images. (a) Standard packaging. (b) Defect packaging.

different welds imperfections have been studied. The results obtained showed that the contamination of metal or sand within the weld joint of two plastic sheets could be clearly identified. Furthermore, THz-TDS allowed to distinguish between welded and non-welded material.

Analysis regarding paper packaging defects have been carried out by the authors of this chapter using THz FiCo system introduced in Section 4.1. The experiments have regarded non destructive testing of sugar bags with and without packaging defects. A standard packaging is reported in **Figure 9a**, while the sachet of sugar in **Figure 9b** is characterized by a surface defect of about 10 mm. The data has been gathered with a 0.4 mm spatial resolution along the x and y axes. **Figure 10** indicates the ability of THz imaging to represent accurately the surface of the packaging. In particular, the defect of the package has been successfully checked guaranteeing an effective control of quality as **Figure 10b** shows. These results allow us to appreciate the diagnostic capabilities of THz technology of detecting packaging defect that could compromise the quality of the food.

5. Challenges and perspectives

THz imaging has interesting features that make it particularly attractive for food quality control. However, there are several issues to be faced in order to move from laboratories and academic environments to a widespread industrial use. In this frame, main issues are:

1. cost of detectors and sources, that is still too high to make THz technology economically convenient for many applications;

2. scanning time, which is not compliant with the time of an industrial production line;

3. high sensitivity to the environmental parameters, which implies a Signal to Noise ratio not suitable for an accurate imaging.

Thanks to technological advancements regarding the development of laser sources, integrated optics and industrial-grade THz hardware, a significant cost reduction has been registered in the last years and it is expected in the future, thus the availability of low-cost and high performance THz imaging systems appears as a reachable goal. Furthermore, high performance multipixel THz cameras look like a valuable solution to perform an accurate real time imaging satisfying someone of the most relevant inspection requests. In this regard, the use of 2D arrays of emitters and detectors is worth to be considered in order to avoid mechanical movement of the sample. Alternatively, single-pixel detection combined with compressive sensing techniques can be employed [56]. There are also challenges related to attenuation, penetration depth and scattering effect of THz radiation to be considered. THz waves suffer a strong absorption by water and appears not effective to inspect high humidity products with a thickness greater than 1 *mm*. In this respect, broadband systems or systems working in the low-frequency THz region are subject of research activities [57]. On the other hand, it is important to study THz interaction with food materials and how it depends on the water type and its level. On the other hand, THz waves exhibit a low penetration depth especially when measuring high-moisture foods (e.g meat; fruit). The penetration depth depends on several factors such as properties of the food material, thickness of the sample, power of THz source, probing wavelength, view and observation angles under which the food is surveyed [58]. Modern THz systems mitigate this problem enhancing the power of the THz radiation [59]. The scattering effect affects both transmission and reflection THz imaging and it is ascribable to in-homogeneity of food, the presence of particles having irregular shape and size comparable to the probing wavelength [25]. This challenges can be faced by using data processing algorithms exploiting an advanced modeling of the interaction between THz radiation and food [25].

Finally, it is worth pointing out that, due to the diversity of samples to be analyzed, most of the high-speed THz systems are application-specific. In fact, in order to improve the performances of THz system in a manufacturing and processing line, the system must be specifically optimized for that particular product and purpose. This process is usually long and labor-intensive.

6. Conclusions and future trends

We presented an overview of the instruments and sensing principles used for THz imaging and its application for food quality inspection. This chapter has begun by presenting the main adopted THz systems and imaging modalities. In particular, we focus our attention on THz imaging in trasmission mode, THz Time of Flight imaging and THz Camera. Then, examples have been summarized briefly in order to show THz imaging applications to different food samples. It emerged that, THz waves are characterized by the ability to pass through a wide variety of packaging materials allowing us to control the quality of the products packaged in glass, plastic, paper and cardboard. On the other hand, THz waves are able to detect low density material hidden in food products through a non-destructive monitoring. Although the effectiveness of THz imaging has been demonstrated for a large number of issues in food quality control, it is clear that there is still a long way to go before this technology can be applied in industrial processes. In fact, the costs of THz instrumentation are still very high, there are limits in the penetration depth that can be investigated and the process is time consuming. So future research is needed to develop fast and economical THz systems through the implementation of compact and more efficient instrumentation. Another important goal to be achieved is the introduction of a database library for the main food component that can simplify the prediction and detection processes. Despite the limitations, THz imaging is establishing itself as a powerful tool for non-destructive inspection in the food industry and for the next future its use is expected to be considered in a wide range of surveys and become more and more consolidated.

Acknowledgements

This work has been supported by PRIN BEST-Food - Broadband Electromagnetic Sensing Technologies for Food quality and security assessment (grant n. 20179*FLH4A*).

Author details

Sonia Zappia[1,2], Lorenzo Crocco[1] and Ilaria Catapano[1*]

1 Institute for the Electromagnetic Sensing of the Environment, National Research Council of Italy, Naples, Italy

2 Department of Electrical Engineering and Information Technology, "Federico II" University of Naples, Italy

*Address all correspondence to: catapano.i@irea.cnr.it

References

[1] Hildur Einarsdóttir, Monica Jane Emerson, Line Harder Clemmensen, Kai Scherer, Konstantin Willer, Martin Bech, Rasmus Larsen, Bjarne Kjær Ersbøll, and Franz Pfeiffer, "Novelty detection of foreign objects in food using multi-modal x-ray imaging," Food Control, vol. 67, pp. 39–47, 2016.

[2] Ronald P Haff and Natsuko Toyofuku, "X-ray detection of defects and contaminants in the food industry," *Sensing and Instrumentation for Food Quality and Safety*, vol. 2, no. 4, pp. 262–273, 2008.

[3] Mikkel Schou Nielsen, Torsten Lauridsen, Lars Bager Christensen, and Robert Feidenhans, "X-ray dark-field imaging for detection of foreign bodies in food," *Food Control*, vol. 30, no. 2, pp. 531–535, 2013.

[4] Giaime Ginesu, Daniele D Giusto, Volker Margner, and Peter Meinlschmidt, "Detection of foreign bodies in food by thermal image processing," *IEEE Transactions on Industrial Electronics*, vol. 51, no. 2, pp. 480–490, 2004.

[5] AA Gowen, BK Tiwari, PJ Cullen, K McDonnell, and CP O'Donnell, "Applications of thermal imaging in food quality and safety assessment," *Trends in food science & technology*, vol. 21, no. 4, pp. 190–200, 2010.

[6] Jayani Chandrapala, Christine Oliver, Sandra Kentish, and Muthupandian Ashokkumar, "Ultrasonics in food processing–food quality assurance and food safety," *Trends in Food Science & Technology*, vol. 26, no. 2, pp. 88–98, 2012.

[7] Bosen Zhao, Ying Jiang, Otman A Basir, and Gauri S Mittal, "Foreign body detection in foods using the ultrasound pulse/echo method," *Journal of food quality*, vol. 27, no. 4, pp. 274–288, 2004.

[8] Chun-Chieh Yang, Moon S Kim, Sukwon Kang, Byoung-Kwan Cho, Kuanglin Chao, Alan M Lefcourt, and Diane E Chan, "Red to far-red multispectral fluorescence image fusion for detection of fecal contamination on apples," *Journal of Food Engineering*, vol. 108, no. 2, pp. 312–319, 2012.

[9] Francesca Vipiana, Lorenzo Crocco, and Joe LoVetri, "Electromagnetic imaging and sensing for food quality and safety assessment [guest editorial]," *IEEE Antennas and Propagation Magazine*, vol. 62, no. 5, pp. 16–17, 2020.

[10] Mohammad Asefi, Ian Jeffrey, Joe LoVetri, Colin Gilmore, Paul Card, and Jitendra Paliwal, "Grain bin monitoring via electromagnetic imaging," *Computers and Electronics in Agriculture*, vol. 119, pp. 133–141, 2015.

[11] Foodradar. [Online]. Available: http://www.foodradar.com/e

[12] M Ricci, L Crocco, and F Vipiana, "Microwave imaging device for in-line food inspection," in *2020 14th European Conference on Antennas and Propagation (EuCAP)*. IEEE, 2020, pp. 1–4.

[13] P Knobloch, C Schildknecht, T Kleine-Ostmann, M Koch, S Hoffmann, M Hofmann, E Rehberg, M Sperling, K Donhuijsen, G Hein, et al., "Medical THz imaging: an investigation of histo-pathological samples," *Physics in Medicine & Biology*, vol. 47, no. 21, pp. 3875, 2002.

[14] Danielle M Charron, Katsuhiro Ajito, Jae-Young Kim, and Yuko Ueno, "Chemical mapping of pharmaceutical cocrystals using terahertz spectroscopic imaging," *Analytical Chemistry*, vol. 85, no. 4, pp. 1980–1984, 2013.

[15] John F Federici, Brian Schulkin, Feng Huang, Dale Gary, Robert Barat, Filipe Oliveira, and David Zimdars, "THz

imaging and sensing for security applications—explosives, weapons and drugs," *Semiconductor Science and Technology*, vol. 20, no. 7, pp. S266, 2005.

[16] Ilaria Catapano and Francesco Soldovieri, "A data processing chain for terahertz imaging and its use in artwork diagnostics," *Journal of Infrared, Millimeter, and Terahertz Waves*, vol. 38, no. 4, pp. 518–530, 2017.

[17] Christian Jördens, Frank Rutz, and Martin Koch, "Quality assurance of chocolate products with terahertz imaging," in *European Conference on Non-Destructive Testing*, 2006.

[18] Hee Jun Shin, Sung-Wook Choi, and Gyeongsik Ok, "Qualitative identification of food materials by complex refractive index mapping in the terahertz range," *Food chemistry*, vol. 245, pp. 282–288, 2018.

[19] Yasuyuki Morita, Adrian Dobroiu, Chiko Otani, and Kodo Kawase, "A real-time inspection system using a terahertz technique to detect microleak defects in the seal of flexible plastic packages," *Journal of food protection*, vol. 68, no. 4, pp. 833–837, 2005.

[20] Fabian Friederich, Wolff Von Spiegel, Maris Bauer, Fanzhen Meng, Mark D Thomson, Sebastian Boppel, Alvydas Lisauskas, Bernd Hils, Viktor Krozer, Andreas Keil, et al., "THz active imaging systems with real-time capabilities," *IEEE Transactions on Terahertz Science and Technology*, vol. 1, no. 1, pp. 183–200, 2011.

[21] P Uhd Jepsen, David G Cooke, and Martin Koch, "Terahertz spectroscopy and imaging–modern techniques and applications," *Laser & Photonics Reviews*, vol. 5, no. 1, pp. 124–166, 2011.

[22] Leili Afsah-Hejri, Parvaneh Hajeb, Parsa Ara, and Reza J Ehsani, "A comprehensive review on food applications of terahertz spectroscopy and imaging," *Comprehensive Reviews in Food Science and Food Safety*, vol. 18, no. 5, pp. 1563–1621, 2019.

[23] Reza Safian, Galia Ghazi, and Nafiseh Mohammadian, "Review of photomixing continuous-wave terahertz systems and current application trends in terahertz domain," *Optical Engineering*, vol. 58, no. 11, pp. 110901, 2019.

[24] Daniel M Mittleman, Rune H Jacobsen, and Martin C Nuss, "T-ray imaging," *IEEE Journal of selected topics in quantum electronics*, vol. 2, no. 3, pp. 679–692, 1996.

[25] Jianyuan Qin, Yibin Ying, and Lijuan Xie, "The detection of agricultural products and food using terahertz spectroscopy: a review," *Applied Spectroscopy Reviews*, vol. 48, no. 6, pp. 439–457, 2013.

[26] DM Mittleman, J Cunningham, MC Nuss, and M Geva, "Noncontact semiconductor wafer characterization with the terahertz hall effect," *Applied Physics Letters*, vol. 71, no. 1, pp. 16–18, 1997.

[27] Chen Wang, Ruiyun Zhou, Yuxin Huang, Lijuan Xie, and Yibin Ying, "Terahertz spectroscopic imaging with discriminant analysis for detecting foreign materials among sausages," *Food Control*, vol. 97, pp. 100–104, 2019.

[28] AA Gowen, Créidhe O'Sullivan, and CP O'Donnell, "Terahertz time domain spectroscopy and imaging: Emerging techniques for food process monitoring and quality control," *Trends in Food Science & Technology*, vol. 25, no. 1, pp. 40–46, 2012.

[29] Yuefang Hua and Hongjian Zhang, "Qualitative and quantitative detection of pesticides with terahertz time-domain spectroscopy," *IEEE Transactions on Microwave Theory and Techniques*, vol. 58, no. 7, pp. 2064–2070, 2010.

[30] Xi-Cheng Zhang and Jingzhou Xu, *Introduction to THz wave photonics*, vol. 29, Springer, 2010.

[31] Jun Takayanagi, Hiroki Jinno, Shingo Ichino, Koji Suizu, Masatsugu Yamashita, Toshihiko Ouchi, Shintaro Kasai, Hideyuki Ohtake, Hirohisa Uchida, Norihiko Nishizawa, et al., "High-resolution time-of-flight terahertz tomography using a femtosecond fiber laser," *Optics express*, vol. 17, no. 9, pp. 7533–7539, 2009.

[32] Binbin B Hu and Martin C Nuss, "Imaging with terahertz waves," *Optics letters*, vol. 20, no. 16, pp. 1716–1718, 1995.

[33] Trygve R Sørgård, Arthur D van Rheenen, and Magnus W Haakestad, "Terahertz imaging of composite materials in reflection and transmission mode with a time-domain spectroscopy system," in *Terahertz, RF, Millimeter, and Submillimeter-Wave Technology and Applications IX*. International Society for Optics and Photonics, 2016, vol. 9747, p. 974714.

[34] Justinas Zdanevičius, Maris Bauer, Sebastian Boppel, Vilius Palenskis, Alvydas Lisauskas, Viktor Krozer, and Hartmut G Roskos, "Camera for high-speed THz imaging," *Journal of Infrared, Millimeter, and Terahertz Waves*, vol. 36, no. 10, pp. 986–997, 2015.

[35] Marcel JE Golay, "The theoretical and practical sensitivity of the pneumatic infra-red detector," *Review of Scientific Instruments*, vol. 20, no. 11, pp. 816–820, 1949.

[36] Jun Yang, Shuangchen Ruan, and Min Zhang, "Real-time, continuous-wave terahertz imaging by a pyroelectric camera," *Chinese optics letters*, vol. 6, no. 1, pp. 29–31, 2008.

[37] PL Richards, "Bolometers for infrared and millimeter waves," *Journal of Applied Physics*, vol. 76, no. 1, pp. 1–24, 1994.

[38] Hichem Guerboukha, Kathirvel Nallappan, and Maksim Skorobogatiy, "Toward real-time terahertz imaging," *Advances in Optics and Photonics*, vol. 10, no. 4, pp. 843–938, 2018.

[39] Michael I Dyakonov and Michael S Shur, "Plasma wave electronics: novel terahertz devices using two dimensional electron fluid," *IEEE Transactions on Electron Devices*, vol. 43, no. 10, pp. 1640–1645, 1996.

[40] Richard Al Hadi, Hani Sherry, Janusz Grzyb, Yan Zhao, Wolfgang Forster, Hans M Keller, Andreia Cathelin, Andreas Kaiser, and Ullrich R Pfeiffer, "A 1 k-pixel video camera for 0.7–1.1 terahertz imaging applications in 65-nm cmos," *IEEE Journal of Solid-State Circuits*, vol. 47, no. 12, pp. 2999–3012, 2012.

[41] Alain Bergeron, Marc Terroux, Linda Marchese, Ovidiu Pancrati, Martin Bolduc, and Hubert Jerominek, "Components, concepts, and technologies for useful video rate THz imaging," in *Millimetre Wave and Terahertz Sensors and Technology V*. International Society for Optics and Photonics, 2012, vol. 8544, p. 85440C.

[42] Georgios C Trichopoulos, H Lee Mosbacker, Don Burdette, and Kubilay Sertel, "A broadband focal plane array camera for real-time THz imaging applications," *IEEE Transactions on Antennas and Propagation*, vol. 61, no. 4, pp. 1733–1740, 2013.

[43] Natsuki Nemoto, Natsuki Kanda, Ryo Imai, Kuniaki Konishi, Masaru Miyoshi, Seiji Kurashina, Tokuhito Sasaki, Naoki Oda, and Makoto Kuwata-Gonokami, "High-sensitivity and broadband, real-time terahertz camera incorporating a micro-bolometer array with resonant cavity structure," *IEEE Transactions on Terahertz Science and Technology*, vol. 6, no. 2, pp. 175–182, 2015.

[44] M Herrmann, M Tani, M Watanabe, and K Sakai, "Terahertz imaging of objects in powders," *IEE Proceedings-Optoelectronics*, vol. 149, no. 3, pp. 116–120, 2002.

[45] Gyeongsik Ok, Hyun Jung Kim, Hyang Sook Chun, and Sung-Wook Choi, "Foreign-body detection in dry food using continuous sub-terahertz wave imaging," *Food control*, vol. 42, pp. 284–289, 2014.

[46] Yuying Jiang, Hongyi Ge, and Yuan Zhang, "Detection of foreign bodies in grain with terahertz reflection imaging," *Optik*, vol. 181, pp. 1130–1138, 2019.

[47] Gyeongsik Ok, Kisang Park, Min-Cheol Lim, Hyun-Joo Jang, and Sung-Wook Choi, "140-ghz subwavelength transmission imaging for foreign body inspection in food products," *Journal of Food Engineering*, vol. 221, pp. 124–131, 2018.

[48] Geun-Ju Kim, Jung-Il Kim, Seok-Gy Jeon, Jaehong Kim, Kyung-Kook Park, and Chang-Hyun Oh, "Enhanced continuous-wave terahertz imaging with a horn antenna for food inspection," *Journal of Infrared, Millimeter, and Terahertz Waves*, vol. 33, no. 6, pp. 657–664, 2012.

[49] Gyeongsik Ok, Kisang Park, Hyang Sook Chun, Hyun-Joo Chang, Nari Lee, and Sung-Wook Choi, "High-performance sub-terahertz transmission imaging system for food inspection," *Biomedical optics express*, vol. 6, no. 5, pp. 1929–1941, 2015.

[50] Y. Lee, S. Choi, S. Han, D. Woo, and H. S. Chun, "Detection of foreign bodies in foods using continuous wave terahertz imaging," *Journal of food protection*, vol. 75, no. 1, pp. 179–183, 2012.

[51] Gyeongsik Ok, Sung-Wook Choi, Kyung Hyun Park, and Hyang Sook Chun, "Foreign object detection by sub-terahertz quasi-bessel beam imaging," *Sensors*, vol. 13, no. 1, pp. 71–85, 2013.

[52] Gyeongsik Ok, Hee Jun Shin, Min-Cheol Lim, and Sung-Wook Choi, "Large-scan-area sub-terahertz imaging system for nondestructive food quality inspection," *Food Control*, vol. 96, pp. 383–389, 2019.

[53] TeraSense. [Online]. Available: http://terasense.com/applications/terahertzfood inspection/

[54] I Catapano and F Soldovieri, "THz imaging and spectroscopy: First experiments and preliminary results," in *2015 8th International Workshop on Advanced Ground Penetrating Radar (IWAGPR)*. IEEE, 2015, pp. 1–4.

[55] S Wietzke, C Jördens, N Krumbholz, B Baudrit, M Bastian, and M Koch, "Terahertz imaging: a new non-destructive technique for the quality control of plastic weld joints," *Journal of the European Optical Society-Rapid Publications*, vol. 2, 2007.

[56] Sang-Heum Cho, Sang-Hun Lee, Chan Nam-Gung, Seoung-Jun Oh, Joo-Hiuk Son, Hochong Park, and Chang-Beom Ahn, "Fast terahertz reflection tomography using block-based compressed sensing," *Optics Express*, vol. 19, no. 17, pp. 16401–16409, 2011.

[57] Yu Kadomura, Naoki Yamamoto, and Keisuke Tominaga, "THz dynamics of hydrated phospholipid studied by broadband dielectric spectroscopy," in *2018 43rd International Conference on Infrared, Millimeter, and Terahertz Waves (IRMMW-THz)*. IEEE, 2018, pp. 1–2.

[58] Albert Redo-Sanchez, Norman Laman, Brian Schulkin, and Thomas Tongue, "Review of terahertz technology readiness assessment and applications," *Journal of Infrared,*

Millimeter, and Terahertz Waves, vol. 34, no. 9, pp. 500–518, 2013.

[59] Deniz Turan, Sofia Carolina Corzo García, Enrique Castro-Camus, and Mona Jarrahi, "Terahertz power enhancement by improving metal adhesion layer of plasmonic photoconductive sources," in *2017 Conference on Lasers and Electro-Optics (CLEO). IEEE*, 2017, pp. 1–2.

Chapter 6

Terahertz Sensing Based on Photonic Crystal Fibers

Md. Ahasan Habib, Md. Shamim Anower and Md. Nazmul Islam

Abstract

Photonic-crystal-fiber (PCF) based sensors in the terahertz spectrum have been immensely studied and implemented due to their unique advantages and high sensitivity. At an early stage, conventional and hybrid structured porous core PCF-based sensors were proposed, but the sensitivity was not so high. With the advancement of PCF fabrication technology, hybrid structured hollow-core PCFs have been reported and offer superior sensing characteristics than the previous types. In this chapter, both porous core and hollow-core PCF-based THz sensors are analyzed and the propagation characteristics are explained using terahertz spectrum. Finally, some promising terahertz sensors are studied and compared at the end of this chapter.

Keywords: terahertz, photonic crystal fiber, relative sensitivity, porous & hollow-core PCF

1. Introduction

The electromagnetic signal whose frequency ranges between 0.1 and 10 THz is termed the terahertz radiation band [1–4]. This radiation band has been effectively used in numerous applications, such as sensing, medical imaging, biotechnology, and genetic engineering, security, chemical spectroscopy, diagnosis of different cancerous cells, radar, and astronomy [5–10]. Due to the massive advancement in the optoelectronics sector, numerous types of terahertz sources and terahertz detectors (photoconductive antennas, bolometers, *etc*.) are already available on the market [11–13]. On the contrary, efficient and favorable terahertz waveguides are still under research. The major barrier for the expansion of terahertz waveguides in different applications is the selection of waveguide base material. Metallic waveguides are suitable for microwave signal propagation, but they offer high ohmic loss for higher frequency signals. A circular-shaped metallic waveguide [14] was proposed in 1999, which suffered only ohmic loss for terahertz signals, whereas coplanar and microstrip transmission lines exhibited other types of losses [15]. One of the major drawbacks of this circular waveguide was the strong dispersion near the cut-off frequency. One year later, a rectangular metallic waveguide [16] was demonstrated theoretically and experimentally, which experienced the same type of loss and dispersion problem as the circular metallic waveguide [15]. In 2001, a copper-based parallel plate terahertz waveguide was proposed, which exhibited only ohmic loss but the dispersion was absent for this type of waveguide [17]. After that, Wang and Mittleman first introduced bare metal or Sommerfeld wire as a terahertz waveguide, and they proved that this waveguide offered less loss than the metallic waveguides [18]. However, in metallic waveguides, the ohmic loss was low, but this waveguide experienced fewer light confinement problems for terahertz waves. This problem can be solved by replacing metallic slit waveguides instead of the bare metal waveguide, and it was first introduced in 2007 [19]. Along with the various types of metallic waveguides, some remarkable dielectric waveguides were also proposed by the researchers, which also include those in the terahertz frequency range. The dielectric waveguides were used to transmit higher frequency signals (such as infrared and optical frequencies) as metallic wires experienced higher losses at these frequencies. On the contrary, the dielectric waveguides were subjected to dielectric absorption loss dependent on waveguide frequencies. Dielectric waveguides can be classified into two categories, which are hollow- and solid-core-fiber waveguides. The dielectric waveguides offer better light confinement capability than metallic waveguides, so the dielectric waveguides are more popular for terahertz wave propagation.

2. Photonic crystal fiber

In hollow-core dielectric waveguide, the light is confined by an omnidirectional mirror and the dielectric material is transparent to a high frequency signal. So that, the hollow-core dielectric waveguide eliminates the limitations existing in silica fiber and hollow metallic waveguide. The light-guiding mechanism in PCFs are exactly similar to the hollow-core dielectric waveguide so that the PCFs can be considered as a dielectric waveguide. The invention of photonic crystal fibers (PCFs) has opened a new door for optoelectronics researchers. However, the selection of the base material for PCF is a tough task as most of the materials experience wavelength-dependent attenuation loss for wide-band signals. In the recent past, a huge number of fused silica-based solid core optical waveguides have been theoretically and experimentally investigated due to some excellent characteristics of this material compared to other glasses or plastics [20]. Fused silica has higher tensile strength, higher transparency to light waves, lower absorption loss, higher availability in nature, high melting point, lower dispersion characteristics, and so on, which increase its popularity as an optical waveguide [21]. However, the silica-based fibers are not suitable for the terahertz spectrum as they offer high absorption loss and modal dispersion to terahertz signals. The most commonly used materials for terahertz dielectric waveguides are TOPAS, Zeonex, Teflon, PMMA, *etc.* due to their lower loss coefficients [15]. Among them, TOPAS and Zeonex offer the lowest loss coefficient and highest transparency towards the terahertz spectrum, so that various application dependent fiber waveguides based on those materials have been proposed [22].

PCFs can be classified into three categories on the basis of the core structure, which are solid core PCF [22], porous core PCF [23, 24], and hollow-core PCF [25]. In addition to wave communication functions, PCFs can also be used for optical sensing. To apply a PCF in sensing applications, the sample under test is filled in the core and after that, the electromagnetic wave is injected through the core [26]. The light interacts with the sample, and after analyzing the received signal, the unknown sample can be identified [26]. However, solid core PCF cannot be used in sensing applications as it is not possible to inject analyte inside of the solid core. So, numerous structures based on hollow-core and porous core dielectric fibers have been proposed as liquid or gas sensors [27–31] using the electromagnetic signal at 1.33 μm or 1.55 μm wavelengths. These proposed PCF-based sensors offer relatively high sensitivity and superior guiding characteristics, which are compared with commonly used mechanical sensors and transducers.

3. PCF waveguide parameters

In order to claim a PCF as a good sensor, it must satisfy some criteria, including high relative sensitivity, low confinement loss, and high numerical aperture. The mathematical formulas or expressions are shown and explained, which are generally used to calculate the sensing and guiding characteristics of an optical sensor.

Relative sensitivity is the key parameter of any PCF-based sensor, which is an indication of how much change of an analyte can be sensed or detected by the sensor. Higher relative sensitivity is desirable from any sensor, as the higher the relative sensitivity, the smaller the change that can be identified by the sensor. The mathematical expression to calculate the relative sensitivity (r) is [27–34],

$$r = \frac{n_r}{n_{eff}} \times P\% \tag{1}$$

where n_r and n_{eff} are the effective refractive index of the actual propagating EM wave through the core and the analytes, respectively. In addition, P indicates how much light signal power interacts with the analytes filled in the fiber core. In the case of a PCF-based sensor, the parameter P is known as power fraction, and that can be easily calculated from the following mathematical expression [27],

$$P = \frac{\int_{sample} R_e\left(E_x H_y - E_y H_x\right) dxdy}{\int_{total} R_e\left(E_x H_y - E_y H_x\right) dxdy} \times 100 \tag{2}$$

where E_x, E_y, Hx, and H_y stand for x and y polarized electric field and magnetic field, respectively, when the propagation direction of light is in the z direction. The numerator of Eq. (2) integrates the energy that interacts with the sample under test and the denominator does the same for the complete sensor.

Another important propagation parameter for any kind of optical fiber-based waveguide or sensor is confinement loss. This loss parameter determines the actual fiber length as higher confinement loss restricts the actual fiber length. This parameter indicates the amount of power wastage due to the cladding air holes and it is usually calculated by employing the following expressions [27–29],

$$\alpha_{CL} = 8.686 \times \frac{2\pi f}{c} \operatorname{Im}\left(n_{eff}\right) \tag{3}$$

where f is the operating frequency of the propagating EM signal, c is the speed of light in free space, and Im (n_{eff}) stands for the imaginary part of the effective refractive index. In order to be considered a good sensor, the loss must be low. Till date, no optical sensor with lower confinement loss has been developed.

Along with the confinement loss, another loss is present in all optical sensors, which is called effective material loss (EML), which reflects the information power consumed by the background material of the sensor. However, many authors did not provide this parameter in their articles, but it is very important for the practical implementation or setup of a terahertz sensor. The mathematical equation to quantify the EML of a sensor is [1],

$$\alpha_{eff} = \frac{\left(\varepsilon_0/\mu_0\right)^{1/2}\int_{A_{mat}} n\ \alpha_{mat}|\mathrm{E}|^2 d\mathrm{A}}{2\int_{All} S_Z d\mathrm{A}} \tag{4}$$

where ε_0 and μ_0 indicate the relative permittivity and permeability of the vacuum, n is the refractive index of the guided light, and α_{mat} is the bulk material loss coefficient of the base material of the sensor. Usually, Topas and Zeonex are the most commonly used materials for the terahertz sensor as they provide the lowest loss coefficient in the terahertz regime. However, different terahertz sensors with PMMA, Teflon, *etc*. were also proposed by the researchers in the recent past which offered slightly higher EML than the recently proposed PCF-based sensors.

Another important propagation parameter is birefringence, which is important in the case of sensing applications. Birefringence is the absolute difference between the effective refractive indexes of x and y polarization mode which is expressed by using the following expression [32, 33],

$$B = |n_x - n_y| \tag{5}$$

where n_x and n_y are the effective refractive indexes in the x and y directions, respectively.

In addition, some other propagation characteristics such as bending loss, numerical aperture, V parameter, spot size, beam divergence, nonlinearity, *etc.* are discussed in [32–45].

4. Photonic-crystal-fiber-based terahertz sensor

A large number of PCF-based terahertz sensors have been proposed in the recent past for the identification of different types of chemicals [32–34], bane chemicals [41], blood components [48–50], alcohols [36], and so on. The prime designing objective of any kind of PCF-based sensor is to propagate the maximum amount of light through the core so that maximum light-analyte interaction takes place. The cross-sectional view of some excellent PCF-based sensors in the terahertz regime is represented in **Figure 1**.

A kagome lattice slotted core fiber is shown in **Figure 1(a)**, which was proposed by Islam *et al.* to detect benzene (n = 1.366), ethanol (n = 1.354), and water (n = 1.33) by using the terahertz spectrum [33]. The kagome structure in **Figure 1(a)** offers high confinement of light through the core due to the compact structure in the cladding. That's the reason why the proposed sensor in **Figure 1(a)** has high relative sensitivity in **Figure 2(a)** and **(b)** for the x and y polarization mode. As the core is not symmetrical, the power distribution in both polarization is unequal.

Another hybrid structured PCF with all rectangular holes is proposed by Islam *et al.* in order to detect three different liquids [37]. Its cross-section is shown in **Figure 1(b)** with an enlarged view of the core section. **Figure 1(b)** shows that the structure is not symmetric, performing birefringence and unequal power distribution in two polarization directions. The relative sensitivity is shown in **Figure 2(c)** and **(d)** for two different core dimensions. As the W value is larger than the H value, more light travels with the x-polarization mode. The sensitivity of a x-polarization mode is thus high. Similarly, **Figure 2(d)** shows the relative sensitivity increases as enlarging the fiber core.

Habib *et al.* proposed a simple hollow-core PCF-based chemical detector [40] using terahertz spectrum and the two-dimensional view is represented in

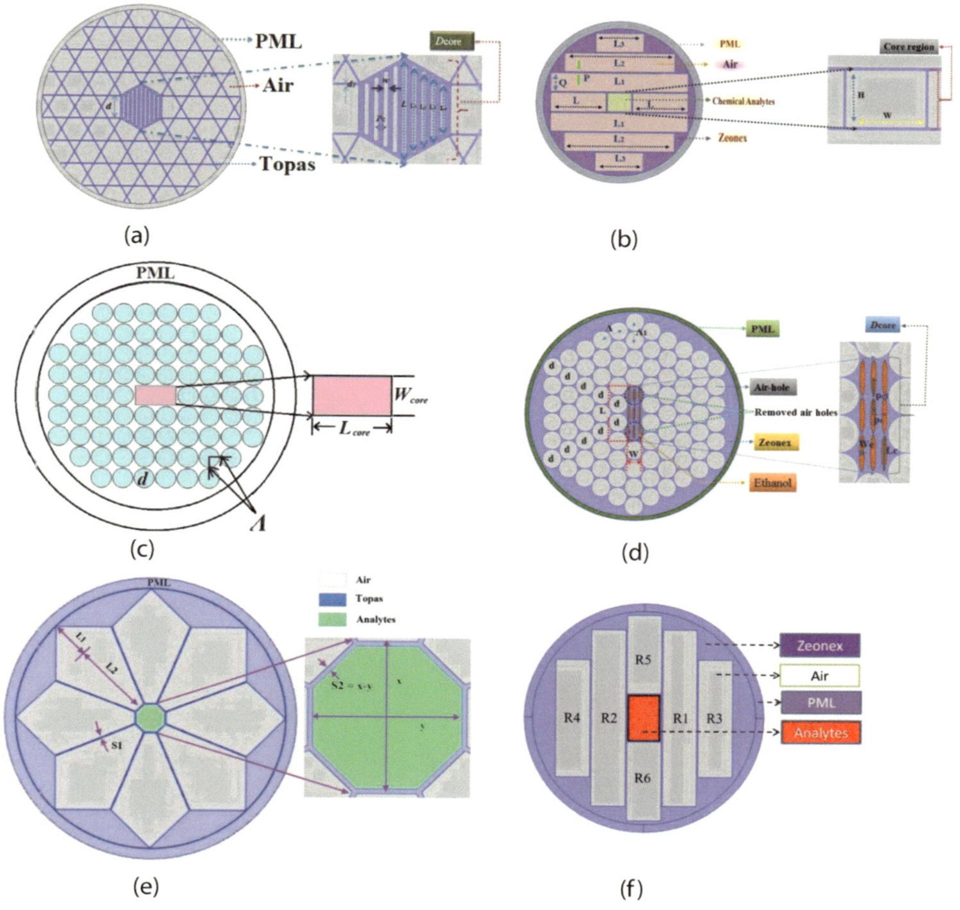

Figure 1.
Cross-sectional view of (a) kagome structured slotted core chemical sensor [33], (b) hybrid structured hollow-core chemical sensor [37], (c) hybrid structured rectangular hollow-core chemical sensor [40], (d) hexagonal structured rectangular porous core alcohol sensor [36], (e) hybrid structured polygonal hollow-core cholesterol sensor [45] and (f) hybrid structured rectangular hollow-core blood component sensor [48] in the terahertz regime.

Figure 1(c). One rectangular air hole is introduced in the core region to be injected by analytes. Due to asymmetric structure, this sensor also has birefringence and the relative sensitivity for both polarization modes are presented in **Figure 2(e)** and **(f)**. This sensor has extremely high sensitivity around 90% at 1.9 THz for the y-polarization mode.

A very simple hollow-core chemical sensor [36] was proposed in 2018 in **Figure 1(d)**. Due to the asymmetric core, the relative sensitivity of that sensor for both polarization modes is reported in **Figure 2(g)** and **(h)**. **Figure 1(d)** indicates that there are considerable amounts of high indexed solid material in the core region. This proposed sensor, therefore, extracts a small fractional THz signal for sensing and results in lower sensitivity. The maximum relative sensitivity is around 73% at 1 THz.

Figure 1(e) represents a hybrid structured hollow-core PCF [45] for the identification of cholesterol present in human blood and fruits. The proposed sensor offered a very high relative sensitivity of 98% at optimum conditions for two main reasons. Firstly, the hollow-core permits maximum light to travel through the core,

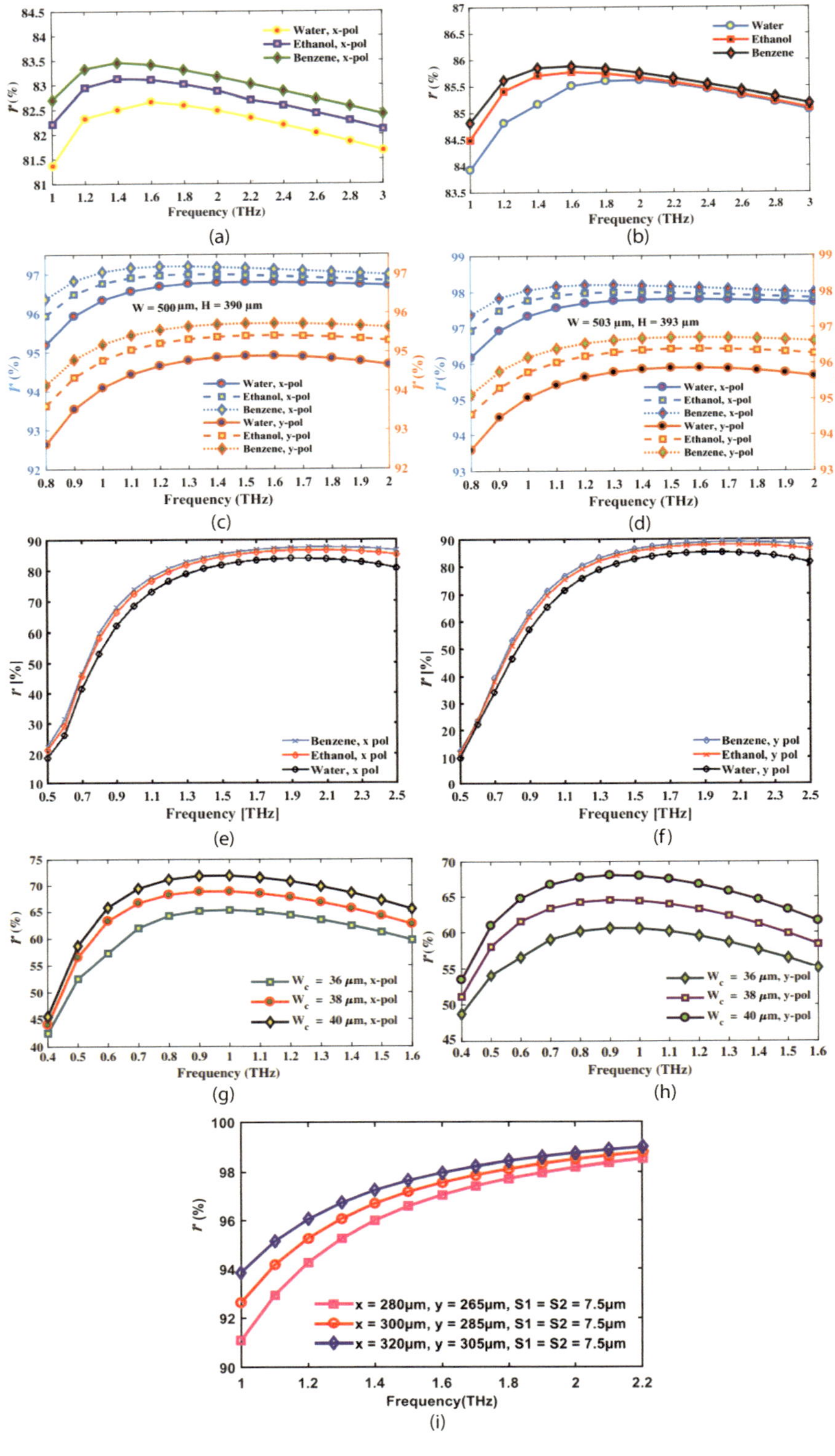

Figure 2.
Relative sensitivity of (a)-(b) kagome structured slotted core chemical sensor [33], (c)-(d) hybrid structured rectangular porous core chemical sensor [37], (e)-(f) hybrid structured rectangular hollow-core chemical sensor [40], (g)-(h) hexagonal structured rectangular porous core alcohol sensor [36], (i) hybrid structured polygonal hollow-core cholesterol sensor [45] for different operating frequencies in the terahertz regime.

which increases the power fraction, and finally, the higher refractive index cholesterol (n = 1.525) attracts more light to interact with itself. So that, according to the data of **Table 1**, the relative sensitivity of this cholesterol sensor is maximum and the graphical representation of the variation of relative sensitivity is shown in **Figure 2(g)**. This figure also shows the relationship between the core dimension and sensitivity. For example, the maximum core dimension has the highest relative sensitivity than at 1.2 THz when the enlarged core has strong light-analyte interaction.

Few PCF-based sensors can identify different components of human blood [48–50]. In 2019, Ahmed *et al*. proposed a hollow-core Zeonex-based blood component sensor in the terahertz spectrum [48], whose cross-sectional view is represented in **Figure 1(f)**. The relative sensitivity of red blood cells (RBCs), hemoglobin, white blood cells (WBCs), plasma, and water for different operating frequencies is shown in **Figure 3(a)** and **(b)**. From the above discussion, it is clear that the relative sensitivity of the terahertz sensors is higher than the silica-based sensors using IR spectrum [27–31]. In addition, hollow-core terahertz sensors offer higher sensitivity (**Figure 2(c)–(f), (i)**) than that of porous core sensors (**Figure 2(a), (b), (g), (h)**).

Now, the loss characteristics of different types of terahertz PCF sensors are discussed below. To reduce the confinement loss and the effective material loss, numerous types of core and cladding structured PCFs were presented [26, 32–35]. However, the light confinement inside the core is largely dependent on the geometric structure of PCF. Now, the variation of confinement loss of PCF-based sensors (**Figure 1**) for different THz frequencies is shown in **Figure 4(a)–(f)**. The first three figures (**Figure 4(a)–(c)**) represent the confinement loss characteristics at certain PCF structures for sensing benzene, ethanol, and water [33, 37, 40]. At high THz frequency, the confinement loss is lower for all analytes because a few fractional THz wave at high frequency travels through the cladding air holes. The confinement loss of benzene is lowest than the other two samples due to the highest refractive index. The sensing results of **Figure 1(d)** and **(e)** are represented in **Figure 4(d)** and **(e)** respectively. The two figures inform that the loss is inversely proportional to the core dimension of the PCFs. Finally, **Figure 4(f)** shows the confinement loss for sensing blood samples [48]. The confinement loss is low at the high THz frequency and high refractive index conditions. THz waves with high frequencies have the tendency to travel through the high indexed zone, so less fractional power propagates through the cladding. The confinement loss is low at high THz frequencies for all PCF-based sensors. The confinement loss is minimum for the cholesterol sensor (**Figure 1(e)**).

Now, the THz wave loss characteristics of different PCF-based sensors in the terahertz regime are investigated. The THz loss spectra are represented in **Figure 5** for different proposed terahertz sensors shown in **Figure 1(a)–(f)**. This loss comes from the solid materials around the core. The THz wave loss characteristics of benzene, ethanol, and water sensors in **Figure 1(b)** and **(c)** are shown in **Figure 5(a)** and **(b)** for different operating frequencies. However, according to **Figure 5(a)** and **(b)**, it is clear that the loss shows an upward trend with the increase of operating frequency. Because THz waves at higher frequency can travel through high indexed solid material, the material more absorbs THz wave energy. **Figure 5(c)** and **(d)** represent THz wave loss characteristics of PCF-based sensors in **Figure 1(d)** and **(e)** for different structural conditions. **Figure 5(c)** shows that THz wave loss is lower for increasing the width of elliptical air holes (**Figure 1(e)**). **Figure 5(d)** shows that the THz wave loss is lowest for the largest core dimension. This structure in **Figure 5(d)** performs the lowest loss among **Figure 5(a)–(d)**.

Ref.	Sensor's structure	Sample's name	r[%]	α_{CL}[dB/m]	α_{eff} [cm^{-1}]	B
[32]	Cladding- Hexagonal Core- Hexagonal Material: Topas	Ethanol	73.5	N/A	N/A	0.011
		Water	71.5	N/A	N/A	0.013
[33]	Cladding- kagome Core- slotted Material: Topas	Benzene	85.9	1.02×10^{-9}	N/A	0.0065
		Ethanol	85.7	1.7×10^{-9}	N/A	0.005
		Water	85.6	4.5×10^{-9}	N/A	0.0043
[34]	Cladding- Hybrid Core- circular Material: Topas	Benzene	78.8	5.83×10^{-9}	N/A	N/A
		Ethanol	78.5	5.81×10^{-10}	N/A	N/A
		Water	69.7	5.34×10^{-8}	N/A	N/A
[35]	Cladding- Hybrid Core- Hybrid Material: Zeonex	HCN	77.5	4.34×10^{-8}	N/A	0.049
		KCN	85.7	4.34×10^{-8}	N/A	0.042
		NACN	87.6	4.34×10^{-8}	N/A	0038
[36]	Cladding- Hexagonal Core- Rectangular Material: Zeonex	Ethanol	68.87	2.66×10^{-9}	0.05	0.0176
[37]	Cladding- Slotted Core- Rectangular Material: Zeonex	Benzene	97.2	1.5×10^{-11}	N/A	0.019
		Ethanol	96.97	3.02×10^{-11}	N/A	0.017
		Water	96.6	2.7×10^{-12}	N/A	0.015
[38]	Cladding- circular Core- circular Material: Silica	Benzene	77.16	1.39×10^{-7}	N/A	N/A
		Ethanol	76.44	1.43×10^{-7}	N/A	N/A
		Water	73.20	1.49×10^{-7}	N/A	N/A
[39]	Cladding-Heptagonal Core- rotated hexa Material: Topas	Benzene	63.24	10^{-10}	N/A	N/A
		Ethanol	61.05	10^{-10}	N/A	N/A
		Water	60.03	10^{-10}	N/A	N/A
[40]	Cladding- Hybrid Core- rectangular hollow Material: Zeonex	Benzene	89	1.15×10^{-7}	N/A	0.007
		Ethanol	88	1.15×10^{-7}	N/A	0.007
		Water	86	1.15×10^{-7}	N/A	0.007
[41]	Cladding- Hexagonal Core- quad elliptical Material: Zeonex	Tabun	63.7	4.34×10^{-5}	0.033	N/A
		Sarin	64.4	4.34×10^{-5}	0.028	N/A
[42]	Cladding- Hybrid Core- rectangular Material: Zeonex	Tabun	95.5	7.42×10^{-12}	0.0094	0.006
		Soman	94.87	9.33×10^{-12}	0.0089	0.00645
		Sarin	94.28	1.24×10^{-11}	0.0086	0.068
[43]	Cladding-Heptagonal Core- rotated hexa Material: Silica	Benzene	69.20	1.92×10^{-9}	N/A	N/A
		Ethanol	68.48	2.13×10^{-9}	N/A	N/A
		Water	66.7	2.7×10^{-6}	N/A	N/A
[44]	Cladding- Hexagonal Core- rotated hexa Material: Topas	Benzene	82.26	6×10^{-8}	N/A	N/A
		Ethanol	81.46	5.85×10^{-8}	N/A	N/A
		Water	79.22	5.84×10^{-8}	N/A	N/A
[45]	Cladding- Eight head star Core- octagonal Material: Topas	Cholesterol	98.75	1.34×10^{-17}	0.0008	N/A

Ref.	Sensor's structure	Sample's name	r[%]	α_{CL}[dB/m]	α_{eff} [cm^{-1}]	B
[46]	Cladding- Hybrid Core- slotted Material: Topas	Benzene	93.95	4.08×10^{-9}	0.0125	N/A
		Water	93.70	2.66×10^{-8}	0.0118	N/A
		HCN	93.40	4.08×10^{-9}	0.0100	N/A
		NaCN	9442	3.77×10^{-6}	0.0149	N/A
		Ketamine	94.87	3.97×10^{-7}	0.0175	N/A
		Amphetamine	94.78	3.17×10^{-6}	0.0165	N/A
[47]	Cladding- Octagonal Core- rotated hexa Material: Topas	Benzene	78.06	3.02×10^{-6}	N/A	N/A
		Ethanol	77.14	2.26×10^{-3}	N/A	N/A
		Water	76.11	2.72×10^{-2}	N/A	N/A
[48]	Cladding-Rectangular slotted Core- Rectangular hollow Material: Zeonex	RBC	93.5	1.35×10^{-11}	N/A	N/A
		HB	92.41	2.16×10^{-11}	N/A	N/A
		WBC	91.25	3.26×10^{-11}	N/A	N/A
		Plasma	90.48	3.95×10^{-11}	N/A	N/A
		Water	89.14	5.64×10^{-11}	N/A	N/A
[49]	Cladding- Hybrid Core- Hybrid Material: Topas	RBC	80.93	1.23×10^{-11}	N/A	N/A
		HB	80.56	8.63×10^{-12}	N/A	N/A
		WBC	80.13	4.93×10^{-12}	N/A	N/A
		Plasma	79.91	2.93×10^{-12}	N/A	N/A
		Water	79.39	1.30×10^{-12}	N/A	N/A
[50]	Cladding- Hexagonal Core- Hybrid Material: Zeonex	RBC	83.45	2.91×10^{-13}	N/A	N/A
		HB	81.20	4.05×10^{-13}	N/A	N/A
		WBC	80.78	8.2×10^{-13}	N/A	N/A
		Plasma	79.60	4.92×10^{-12}	N/A	N/A
		Water	78.80	3.49×10^{-12}	N/A	N/A
[51]	Cladding-Hybrid Core-Circular Material: Zeonex	RBC	95.80	3.80×10^{-11}	N/A	N/A
		HB	95	1.13×10^{-11}	N/A	N/A
		WBC	93.6	2.15×10^{-10}	N/A	N/A
		Plasma	92.5	6.25×10^{-10}	N/A	N/A
		Water	91.4	8.30×10^{-9}	N/A	N/A

Table 1.
Comparison of different guiding and sensing parameters for numerous types of remarkable terahertz sensors.

The following comparison shows the guiding characteristics for hybrid structured sensors [34, 35, 40, 42, 46, 48] and conventional structured sensors [33, 36–39, 41, 43–45, 47, 49–51] in terahertz regime.

The most interesting characteristic of PCF-based sensors is that the guiding or sensing properties are highly dependent on the geometry or structure of the PCFs. However, from **Table 1**, it is quite clear that the relative sensitivity and loss profile is lower for nonconventional structured PCF than regular structured one. For example, the conventional structured liquid sensor in Ref. [44] offered the highest sensitivity of 82.26% among regular structured PCFs but the core loss is high (>10 dB^{-8} dB/m). On the other hand, a maximum of 93.5% relative sensitivity with

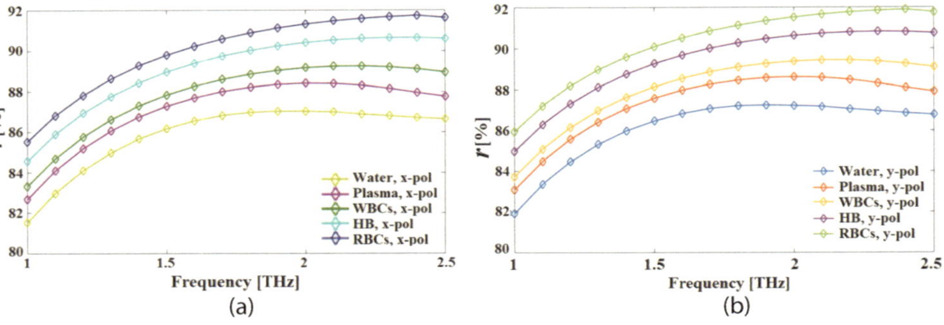

Figure 3.
Relative sensitivity as a function of frequency for different blood components in terahertz regime from (a)-(b) in ref. [48].

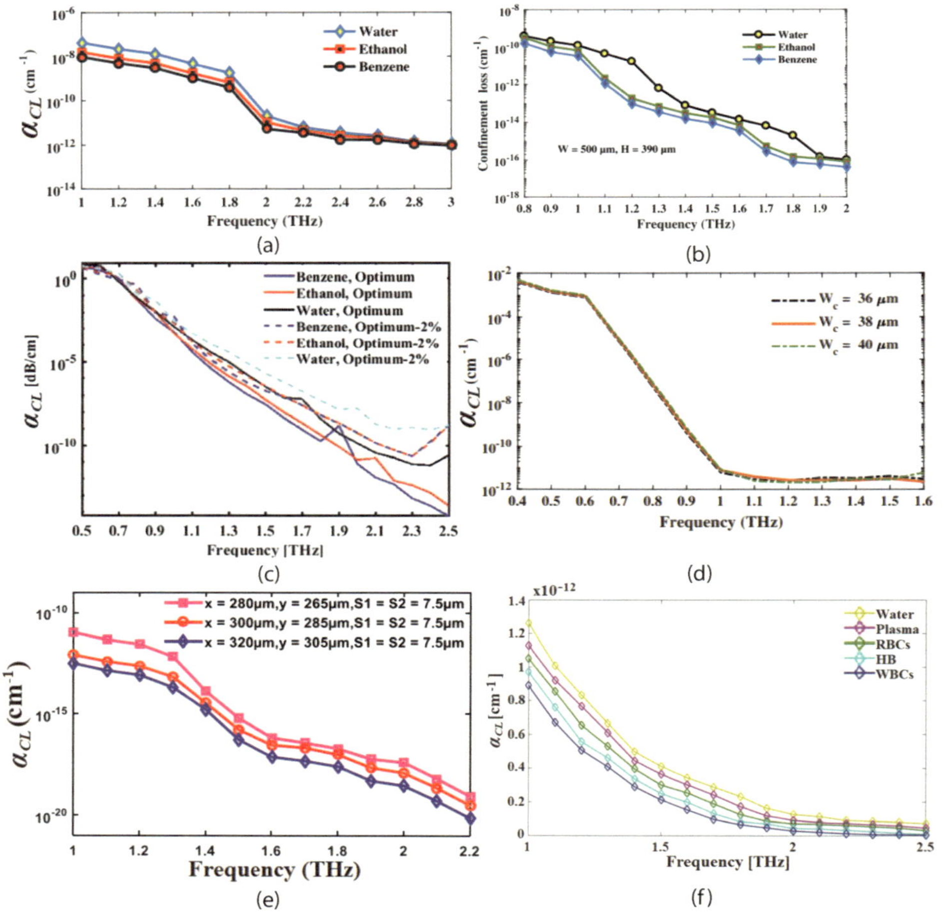

Figure 4.
Confinement loss as a function of frequency of (a) ***Figure 1(a)*** *ref. [33], (b)* ***Figure 1(b)*** *ref. [37], (c)* ***Figure 1(c)*** *ref. [40], (d)* ***Figure 1(d)*** *ref. [36], (e)* ***Figure 1(e)*** *ref. [45] and (f)* ***Figure 1(f)*** *ref. [48] for different PCF based sensors.*

lower confinement loss (< 10 dB^{-9} dB/m) can be achieved from PCF-based sensor in Ref. [46]. Moreover, the relative sensitivity and loss are also dependent on the test sample also. In **Table 1** the highest relative sensitivity (98.75%) and lowest loss

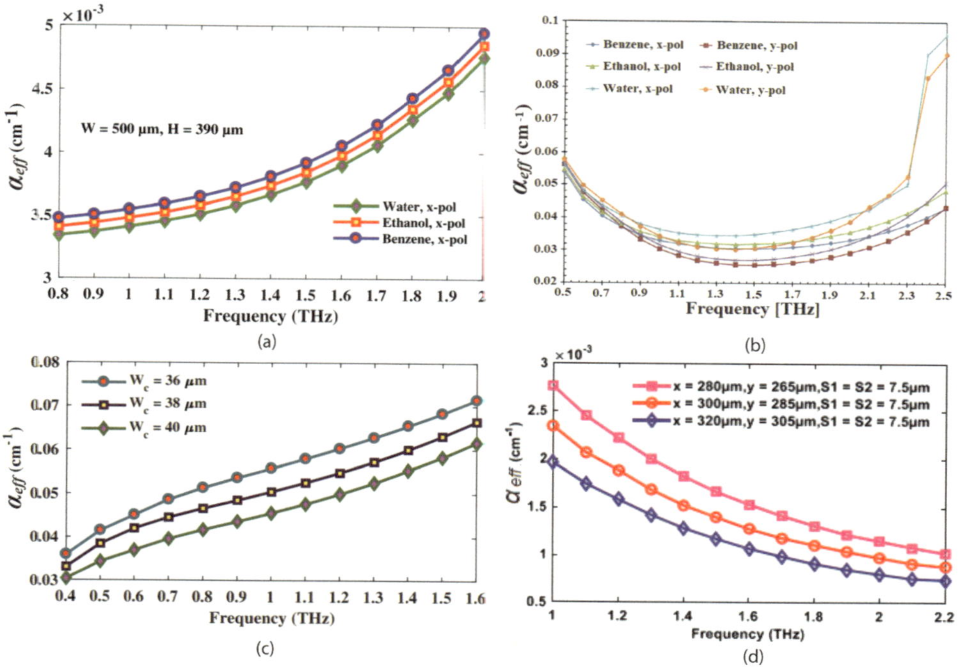

Figure 5.
Effective material loss variation of numerous structured terahertz sensorsas a function of frequency of (a) ***Figure 1(b)*** *ref. [37], (b)* ***Figure 1(c)*** *ref. [40], (c)* ***Figure 1(d)*** *ref. [36], (d)* ***Figure 1(e)*** *ref. [45].*

(core loss = 1.34 × 10 dB^{-17} dB/m and THz wave loss = 0.008 cm^{-1}) was reported in Ref. [45] as the refractive index of cholesterol (refractive index = 1.525) is maximum than any other samples reported in **Table 1**.

5. Conclusion

In this chapter, different types of PCF-based terahertz sensors were discussed and compared for numerous chemicals, toxic agents, and blood components identification. In addition, a brief discussion about terahertz technology and different terahertz waveguides are included in this chapter. The sensing and guiding parameters of numerous terahertz sensors are numerically discussed and graphically represented for a better understanding of the readers. However, efficient application-specific terahertz sensors are still under research and we hope this chapter will help the optoelectronics researchers to propose new sensors.

Author details

Md. Ahasan Habib[1*], Md. Shamim Anower[1] and Md. Nazmul Islam[2]

1 Department of Electrical and Electronic Engineering, Rajshahi University of Engineering and Technology, Rajshahi, Bangladesh

2 Department of Electrical and Electronic Engineering, Bangabandhu Sheikh Mujibur Rahman Science and Technology University, Gopalganj, Bangladesh

*Address all correspondence to: habib.eee.116.ah@gmail.com

References

[1] Ahlbom A, Bridges J, Seze R, Hillert L, Juutilainen J, Mattsson M, et al. Possible effects of electromagnetic fields (EMF) on human health—Opinion of the Scientific Committee on Emerging and Newly Identified Health Risks (SCENIHR). Toxicology. 2008;**246**:248-250

[2] Yang X, Zhao X, Yang K, Liu Y, Fu W, Luo Y. Biomedical applications of terahertz spectroscopy and imaging. Trends in Biotechnology. 2016;**34**(10): 810-824

[3] Habib MA, Anower MS. Design and numerical analysis of highly birefringent single mode fiber in THz regime. Optical Fiber Technology. 2009;**47**:197-203

[4] Reza MS, Habib MA. Extremely sensitive chemical sensor for terahertz regime based on a hollow core photonic crystal fiber. Ukrainian Journal of Physical Optics. 2020;**21**(1):8-14

[5] Zhang XC. Terahertz wave imaging: Horizons and hurdles. Physics in Medicine & Biology. 2002;**47**(21):3663

[6] Wu K, Qi C, Zhu Z, Wang C, Song B, Chang C. Terahertz wave accelerates DNA unwinding: A molecular dynamics simulation study. Journal of Physical Chemistry Letters. 2020;**11**(17):7002-7008

[7] Song Q, Zhao Y, Redo-Sanchez A, Zhang C, Liu X. Fast continuous terahertz wave imaging system for security. Optics Communications. 2009; **282**(10):2019-2022

[8] Zhang Z, Ding H, Yang X, Liang L, Wei D, Wang M, et al. Sensitive detection of cancer cell apoptosis based on the non-bianisotropicmetamaterial biosensors in terahertz frequency. Optical Material Express. 2018;**8**(3):659-667

[9] Cheon H, Paik JH, Choi M, Yang H, Son J. Detection and manipulation of methylation in blood cancer DNA using terahertz radiation. Scientific Reports. 2019;**9**(6413)

[10] Cooper KB, Dengler RJ, Liombart N, Bryllrt T, Chattopadhyay G, Mehdi I, et al. An approach for sub-second imaging of concealed objects using terahertz radar. Journal of Infrared Millimeter, Terahertz and Waves. 2009;**30**:1297-1307

[11] Shrekenhamer D, Watts C, Padilla W. Terahertz single pixel imaging with an optically controlled dynamic spatial light modulator. Optics Express. 2013;**21**(10): 12507-12518

[12] Lewis RA. A review of terahertz sources. Journal of Physics: D. 2014;**47**: 374001

[13] Yardimci NT, Cakmakyapan S, Hemmati S, Jarrahi M. A high power broadband terahertz source enabled by three dimensional light confinement in a plasmonic nanocavity. Scientific Reports. 2017;**7**:4166

[14] McGowan RW, Gallot G, Grischkowsky D. Propagation of ultra-wideband short pulses of THz radiation through sub millimeter-diameter circular waveguides. Optics Letters. 1999;**24**:1431-1433

[15] Atakaramians S. Terahertz waveguides: A study of microwires and porous fibers. Available from: http://hdl.handle.net/2440/69317. 2011

[16] Gallot G, Jamison SP, McGowan RW, Grischowsky D. Terahertz waveguides. Journal of Optical Society of America B. 2000;**17**(5):851-863

[17] Mendis R, Grischkowsky D. Undistorted guided wave propagation of sub picosecond terahertz pulses. Optics Express. 2001;**26**(11):846-848

[18] Wang K, Mittleman DM. Metal wires for terahertz wave guiding. Nature. 2004;**432**:376-379

[19] Wachter M, Nagel M, Kurz H. Metallic slit waveguide for dispersion free low loss terahertz signal transmission. Applied Physics Letter. 2007;**90**:061111

[20] Kuhlmey BT, Eggleton BJ, Wu DKC. Fluid filled solid-core photonic bandgap fibers. Journal of Lightwave Technology. 2009;**27**(11):1617-1630

[21] You J, Lin S, Zhang J, Lin S, Fu L, Zheng R, et al. Optical properties of He^{+} implanted fused silica glass waveguides. International Journal of Modern Physics B. 2021;**35**(2):2150026

[22] Sterke CM, Grujic T, Kuhlmey BT, Argyros A. Solid core photonic crystal fiber with ultrawide bandgap. Frontiers in Optics. 2020. DOI: 10.1364/FIO.2010.FTuW2

[23] Habib MA, Anower MS, Hasan MR. Highly birefringent and low effective material loss microstructure fiber for terahertz wave guidance. Optics Communications. 2018;**423**:140-144

[24] Habib MA, Anower MS. Square porous core microstructure fiber for low loss terahertz applications. Optics and Spectroscopy. 2019;**126**(5):607-613

[25] Couny F, Benabid F, Light PS. Large pitch kagome structured hollow core photonic crystal fiber. Optics Letters. 2006;**31**(24):3574-3576

[26] Habib MA, Vera ER, Aristizabal JV, Anower MS. Numerical modelling of a rectangular hollow core waveguide for the detection of fuel adulteration in terahertz region. Fibers. 2020;**8**:63

[27] Asaduzzaman S, Ahmed K, Bhuiyan T, Farah T. Hybrid photonic crystal fiber in chemical sensing. Springer Plus. 2016;**5**:748

[28] Paul BK, Ahmed K, Asaduzzaman S, Islam MS. Folded cladding porous shaped photonic crystal fiber with high sensitivity in optical sensing applications: Design and analysis. Sensing and Bio-Sensing Research. 2017; **12**:36-42

[29] Podder E, Hossain MB, Jibon RH, Bulbul AAM, Mondal HS. Chemical sensing through photonic crystal fiber: Sulfuric acid detection. Frontiers in Optoelectronics. DOI: 10.1007/s12200-019-0903-8

[30] Leon MJBM, Abedin S, Kabir MA. A photonic crystal fiber for liquid sensing applications with high sensitivity, birefringence and low confinement loss. Sensors International. DOI: 10.1016/j.sintl.2020.100061

[31] Eid MMA, Habib MA, Anower MS, Rashed ANZ. Highly sensitive nonlinear photonic crystal fiber based sensor for chemical sensing applications. Microsystem Technologies. DOI: 10.1007/s00542-020-05019-w

[32] Rana S, Kandadai N, Subbaraman H. A highly sensitive polarization maintaining photonic crystal fiber sensor operating in the THz regime. Photonics. 2018;**5**(40):1-9

[33] Islam MS, Sultana J, Ahmed K, Islam MR, Dinovitser A, Ng BW, et al. A novel approach for spectroscopic chemical identification using photonic crystal fiber in the terahertz regime. IEEE Sensors Journal. 2018;**18**(2):575-582

[34] Paul BK, Ahmed K, Vigneswaran D, Ahmed F, Roy S, Abbott D. Quasi photonic crystal fiber based spectroscopic chemical sensor in the Terahertz spectrum: Design and analysis. IEEE Sensors Journal. 2018; **18**(24):9948-9954

[35] Islam MS, Sultana J, Dinovitser A, Ng BW, Abbott D. Sensing of toxic chemicals using polarized photonic crystal fiber in terahertz regime. Optics Communications. 2018;**426**:341-347

[36] Sultana J, Islam MS, Ahmed K, Dinovitser A, Ng BW, Abbott D. Terahertz detection of alcohol using a photonic crystal fiber sensor. Applied Optics. 2018;**57**(10):2424-2433

[37] Islam MS, Sultana J, Rifat AA, Dinovitser A, Ng BW, Abbott D. Terahertz sensing in a hollow core photonic crystal fiber. IEEE Sensors Journal. 2018;**18**(10):4073-4080

[38] Sen S, Ahmed K. Design of terahertz spectroscopy based optical sensor for chemical detection. SN Applied Sciences. 2019;**1**:1215

[39] Hasan MM, Sen S, Rana MJ, Paul BK, Habib MA, Daiyan GM, et al. Heptagonal photonic crystal fiber based chemical sensor in THz regime. In: Joint 2019 8th International Conference on Informatics, Electronics & Vision (ICIEV) & 3rd International Conference on Imaging, Vision & Pattern Recognition (IVPR).

[40] Habib MA, Anower MS, Abdulrazak LF, Reza MS. Hollow core photonic crystal fiber for chemical identification in terahertz regime. Optical Fiber Technology. 2019;**52**:101933

[41] Yakasai I, Abas PE, Kaijage SF, Caesarendra W, Begu F. Proposal for a quad-elliptical photonic crystal fiber for terahertz wave guidance and sensing chemical warfare liquids. Photonics. 2019; **6**:78

[42] Hossain MB, Podder E, Bulbul AA, Mondal HS. Bane chemical detection through photonic crystal fiber in terahertz regime. Optical Fiber Technology. 2020;**54**:102102

[43] Hossain MS, Sen S. Design and performance improvement of optical chemical sensor based photonic crystal fiber (PCF) in the Terahertz (THz) wave propagation. SILICON. DOI: 10.1007/s12633-020-00696-8

[44] Sen S, Shafi MAA, Kabir MA. Hexagonal photonic crystal Fiber (H-PCF) based optical sensor with high relative sensitivity and low confinement loss for terahertz (THz) regime. Sensing and Bio-Sensing Research. 2020;**30**: 100377

[45] Rahman MM, Mou FA, Bhuiyan MIH, Islam MR. Photonic crystal fiber based terahertz sensor for cholesterol detection in human blood and liquid foodstuffs. Sensing and Bio-Sensing Research. 2020;**29**:100356

[46] Rahman A, Khaleque A, Ali MY, Rahman MT. THz spectroscopic sensing of liquid chemicals using a photonic crystal fiber. OSA Continuum. 2020; **3**(11):2982-2996

[47] Shafi MAA, Sen S. Design and analysis of a chemical sensing octagonal photonic crystal fiber (O-PCF) based optical sensor with high relative sensitivity for terahertz (THz) regime. Sensing and Bio-Sensing Research. 2020;**29**:100372

[48] Hossain MB, Podder E. Design and investigation of PCF-based blood components sensor in terahertz regime. Applied Physics A. 2019;**125**:861

[49] Ahmed K, Ahmed F, Roy S, Paul BK, Akter MN, Vigneswaran D, et al. Refractive index based blood components sensing in Terahertz spectrum. IEEE Sensors Journal. 2019; **19**(9):3368-3375

[50] Habib MA. A refractive index based micro-structured sensor forblood components detection in terahertz regime. Sensor Letters. 2020;**18**(1):74-81

[51] Eid MMA, Habib MA, Anower MS, Rashed ANZ. Hollow core photonic crystal fiber (PCF)–based optical sensor for blood component detection in Terahertz spectrum. Brazilian Journal of Physics. 2021;**51**:1017-1025

Section 3

Terahertz Optoelectronic Technology

Chapter 7

Optical Heterodyne Measurement of Terahertz Wave

Shin'ichiro Hayashi and Norihiko Sekine

Abstract

One of the most notable frequency regions in terms of research currently lies in the 'frequency gap' region between microwaves and infrared: terahertz wave. Although new methods for generating and detecting terahertz wave have been developed, few detectors operating at room temperature are able to capture low-energy terahertz beams. Here we introduce the optical heterodyne measurement (nonlinear frequency up-conversion detection) of terahertz wave using parametric wavelength conversion in a nonlinear crystal; this has better sensitivity than many commonly used thermal detectors such as pyroelectric detectors. Additionally, optical heterodyne techniques allow the beams of terahertz wave to be visualized and their frequency and intensity determined directly as visible light. These are very promising for extending applied researches into the terahertz region, and we expect that these will open new research fields such as wireless information communications or non-destructive inspection in the terahertz region.

Keywords: terahertz wave, nonlinear optics, parametric wavelength conversion

1. Introduction

Terahertz ($10^{\wedge 12}$ Hz) wave region are important not only in the basic sciences, such as molecular optics, molecular spectroscopy, electron acceleration, plasma measurement, and radio astronomy, but also in numerous applications, such as broadband wireless information communication, non-destructive inspection, high precision radar, and global environmental measurement, since they have higher directivity in the atmosphere than microwaves and higher transmittances in soft materials than infrared. Therefore, sensitive, wideband, coherent detectors for terahertz wave that could be widely used in such applications are required. The terahertz wave region is relatively unexplored because of the lack of the commercially available sources, detectors, and optics which have resulted in what is known as the frequency gap [1–3]. Over the past two decades, there has been remarkable growth in the fields of science and engineering using terahertz wave region, which has become a vibrant, international, cross-disciplinary research activity [4]. An effective method for generating and detecting coherent terahertz waves is wavelength conversion in nonlinear optical materials owing to the high conversion efficiency, wide bandwidth, and room-temperature operation. The large figure of merit (FOM) of $LiNbO_3$ at room temperature makes this well-known nonlinear crystal ideal for such applications; terahertz wave parametric generation and detection in $LiNbO_3$ are realized by stimulated polariton scattering via transverse optical phonons [5–10]. In this paper, we introduce the measurement method of terahertz wave by up-conversion. Terahertz wave is generally measured using thermal detectors, such as Si-bolometers, pyroelectric detectors or Golay cells. Although these detectors can measure the average power, the frequency and the phase information are lost. Furthermore, high sensitivity bolometers require ultralow temperature and show slow response. Thus, there is a critical need for frequency and phase measurements, sensitive and fast detectors that work at room temperature, so that terahertz-wave can be used more widely in a variety of important applications.

2. Terahertz wave parametric wavelength conversion

When a strong electric field, such as that of a Q-switched laser pulse, is incident upon a nonlinear optical crystal, the transverse photon and phonon wave fields become coupled and behave as new mixed photon–phonon states, called polaritons. Broadband terahertz wave generation results from efficient parametric wavelength conversion of laser beam via polaritons [5, 6]. The polaritons exhibit phonon-like behavior in the resonant frequency region (near the transverse optical (TO)-phonon frequency ω_{TO}), however, they behave like photons in the non-resonant low-frequency region, as shown in **Figure 1**. Detecting terahertz wave can be achieved by injecting a "seed" for the broadband terahertz wave. Spectroscopic measurement is accomplished simply by observing the propagation direction of idler beams depending on the angle between the incident pumping beam and the wavelength of the seeding terahertz wave. In the parametric wavelength conversion process, a photon of terahertz wave and a photon of near-infrared idler beam are created parametrically from a photon of near-infrared pumping beam, according to the energy conservation law $\omega_p = \omega_T + \omega_i$ (where ω indicates frequency, and p, T, and i denote the pumping beam, terahertz wave, and idler beam, respectively) and the momentum conservation law $\boldsymbol{k_p} = \boldsymbol{k_i} + \boldsymbol{k_T}$ (noncollinear phase-matching condition). This condition leads to the angle-dispersive characteristics of the idler beams and the terahertz waves. Thus, broadband terahertz waves can be detected depending on the phase-matching angle, as shown in **Figure 2**.

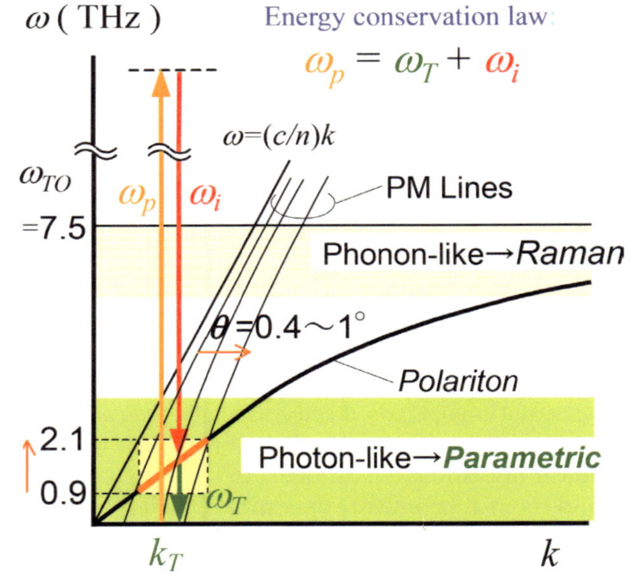

Figure 1.
Dispersion relation of the polariton.

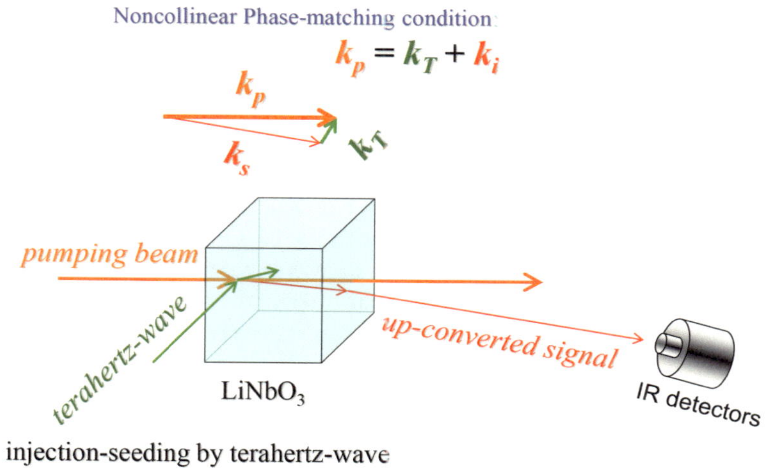

Figure 2.
Noncollinear phase-matching condition.

In our experiment, we used $LiNbO_3$ as nonlinear crystals. The bandwidth of the terahertz wave parametric wavelength conversion is decided by the parametric gain and absorption coefficients in the terahertz region. **Figure 3** shows the calculated gain and the absorption coefficient at the pumping intensity of 0.1, 0.2, 0.4, 0.8, 1.6, and 3.2 GW/cm^2 [11]. As the intensity of pumping beam increases, the gain coefficient also increases in whole frequency region, and peak of the gain curve moves towards higher frequencies. All gain curves have a broad bandwidth with a dip appearing at around 2.6 THz because the low frequency modes of doped MgO in the $LiNbO_3$ work as a crystal lattice defects. The effective parametric gain curve depends on the pumping beam intensity and diameter in noncollinear phase-matching conditions. The idler and terahertz wave are parametrically amplified while propagating coaxially with respect to the pumping wave.

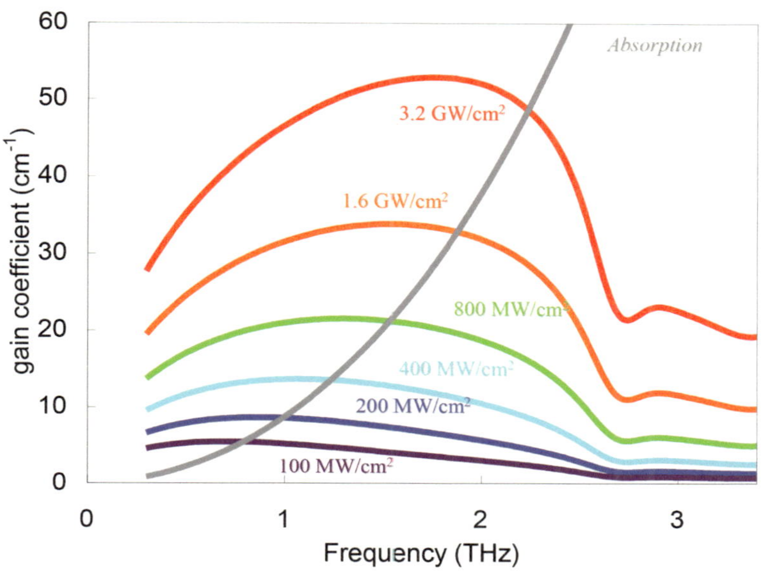

Figure 3.
Calculated gain and absorption coefficient using MgO:$LiNbO_3$ pumped by Nd:YAG laser for several pumping intensities, 0.1, 0.2, 0.4, 0.8, 1.6, and 3.2 GW/cm^2.

3. Experiment

The experimental apparatus for terahertz wave spectroscopic measurement shown in **Figure 4** consists of a nonlinear crystal ($MgO:LiNbO_3$) with a Silicon prism and a laser beam visualizer. The terahertz wave was focused onto the $MgO:LiNbO_3$ crystal via Silicon prism input coupler. The incident angle between the terahertz wave and the pumping beam from Nd:YAG based MOPA system satisfies the noncollinear phase-matching conditions in the $MgO:LiNbO_3$ crystal. Mixing the pumping beam with the incident terahertz wave created a seeded, up-converted (difference-frequency) signal, which was parametrically amplified by the optical parametric amplifier in the $MgO:LiNbO_3$. The up-converted signals were visualized using an infrared laser visualizer (10VIZ-35, Standa Corp.) as visible lights; the frequency and intensity was determined from the position and the intensity.

Figure 5 shows pictures of the up-converted signals on the laser beam visualizer when the input frequency of terahertz wave of 1.0, 1.3, 1.6, 1.9 and 2.2 THz. The scales at the bottom show the calculated distance from the pumping beam. The position and the intensity of the up-converted signal depend on the frequency and the energy of the input terahertz wave respectively. As the frequency of the input terahertz wave increases, the position depending on noncollinear phase-matching angle between the pumping beam and up-converted beam also increases,

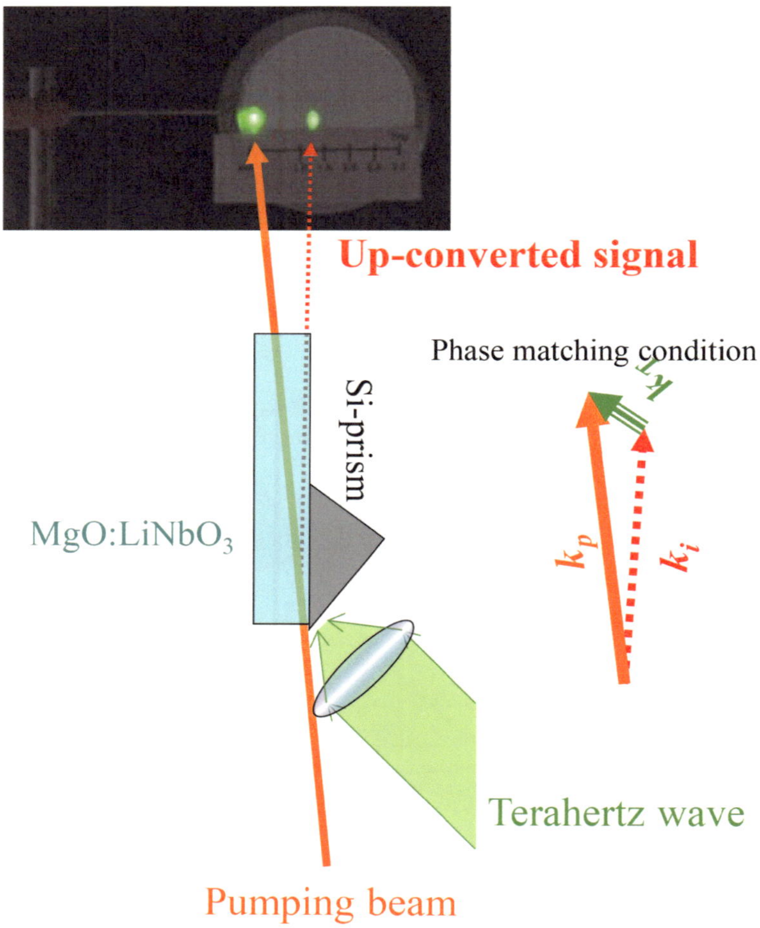

Figure 4.
Experimental apparatus for terahertz wave spectroscopic measurement.

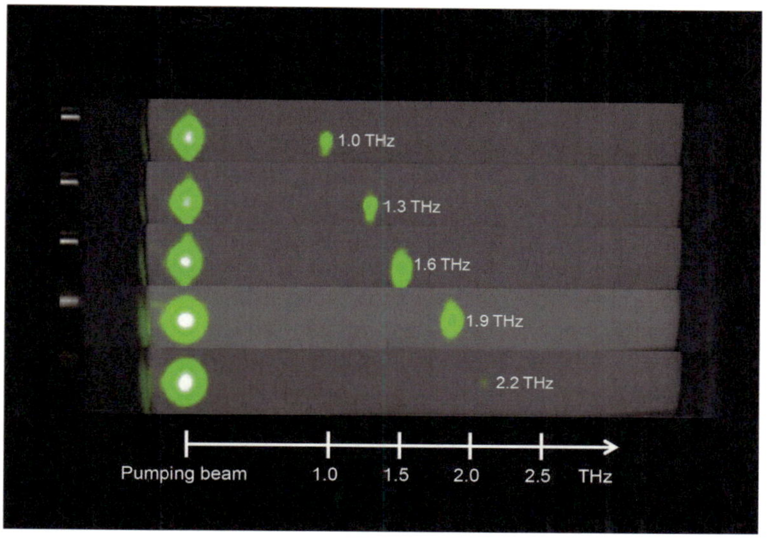

Figure 5.
Pictures of the up-converted signals on the laser beam visualizer when the input frequency of terahertz wave of 1.0, 1.3, 1.6, 1.9 and 2.2 THz.

which shifts the position of up-converted beam. The energy of up-converted beam also depends on the intensity of input terahertz wave, which changes the intensity of the light spot. These make it easy to identify the frequency and intensity of the input terahertz wave from the position and intensity of the up-converted signal. This is in good agreement with the calculation.

Figure 6 shows the input (terahertz wave) – output (up-converted beam) characteristics of optical heterodyne at 1.8 THz when the pumping energy was 10 mJ/pulse. Horizontal axis is input energy of terahertz waves, vertical axis is output energy of the up-converted signals. These colors in the figure represent operating ranges of thermal detectors. The green represents pyroelectric detector at room temperature. The orange represents standard 4 K Si-bolometer. As the input energy of terahertz wave was decreased, the output energy of up-converted signal was also

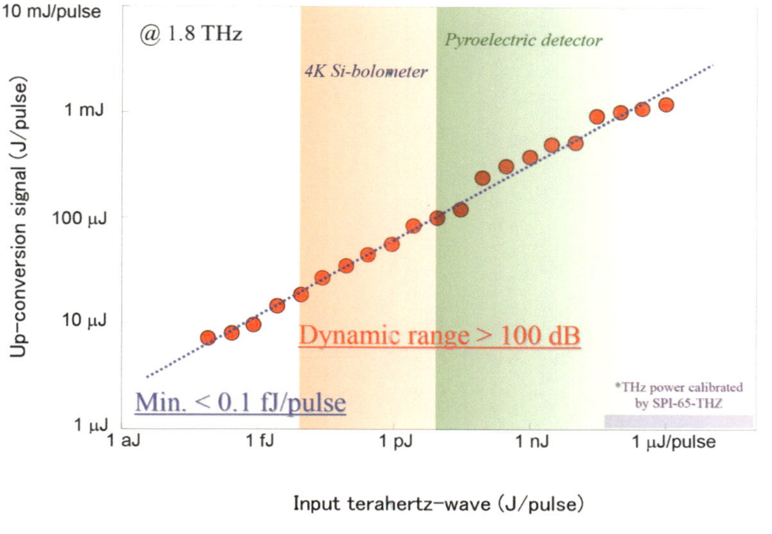

Figure 6.
Output energy of up-converted signals as a function of the input energy of the terahertz wave.

decreased. The maximum energy of up-converted signal was >1 mJ/pulse when the input energy of terahertz-wave was ~1 μJ/pulse. This is easily visualized by laser beam visualizer. The minimum detectable energy of terahertz-wave was less than 100 aJ/pulse. This has higher sensitivity than standard 4 K Si-bolometer. Further, dynamic range reached more than 100 dB.

Figure 7 shows our experimental setup for phase measurement. We used an injection seeded parametric generator (is-TPG) as a terahertz wave source. The terahertz wave from an is-TPG was collimated, passed through a delay line (a rooftop mirror on a mechanical stage). The terahertz wave separated by a wire grid polarizer focused to the $MgO:LiNbO_3$ crystal for up-conversion detection. The incident angle between the pumping beam, the seeded idler beam, and the input terahertz wave satisfy the noncollinear phase-matching conditions in the $MgO:LiNbO_3$ crystal. Mixing the intense pumping beam and the seeded idler beam with the terahertz wave created an up-converted signal. In this case, the intensity of up-converted signal depends on the phase difference between the seeded idler beam and the terahertz wave.

Figure 8 shows the energy of up-converted signals as a function of path length difference between the terahertz wave and the seeded idler wave when the input energy of the pumping beam, the terahertz wave and the idler beam were 10 mJ/pulse, 1 μJ/pulse and 1 μJ/pulse, respectively. As the optical pass length difference increases, we observed blinking of up-converted signals due to the interference between the seeded idler beam and the terahertz wave. The intervals of fringe represent the wavelength of input terahertz-wave. The observed intervals of fringe corresponding to the wavelength of terahertz wave are 200 μm. These show excellent agreement with the wavelength of input terahertz wave from the is-TPG.

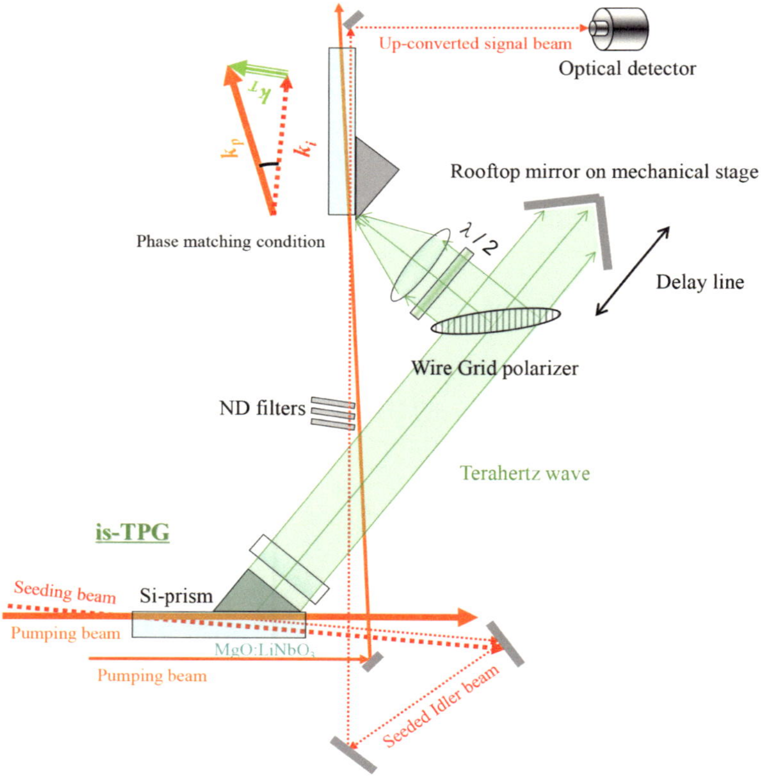

Figure 7.
Experimental apparatus for terahertz wave phase measurement.

DOI: http://dx.doi.org/10.5772/intechopen.99168

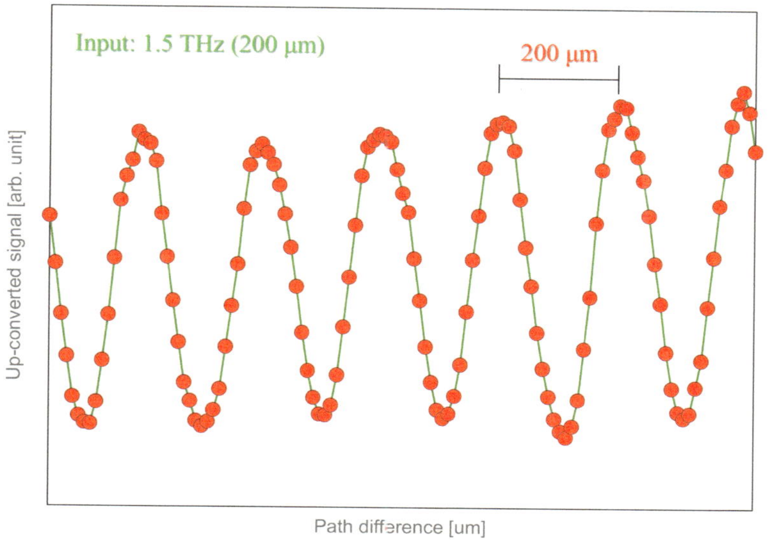

Figure 8.
Phase measurement of terahertz wave by up-conversion.

4. Conclusion

We have introduced here spectroscopic, sensitive, coherent, wideband terahertz wave optical heterodyne detection via wavelength conversion in MgO:$LiNbO_3$. The coherent detection is very important for the calibration of terahertz wave frequency. In general, the radiation frequency calibration is based on absorption lines of molecular gases as a traceable standard, but there is no radiation frequency standard in the terahertz region [12]. In the case, the spectroscopic information of terahertz wave is measured by well-developed optical measurement methods as a traceable standard. Under the experimental conditions, the observed minimum energy of terahertz wave is less than 100 aJ/pulse. This is much higher sensitivity than many cryogenic thermal detectors. Furthermore, the sensitivity depending on parametric gain of the terahertz wave in $LiNbO_3$ could be higher by cooling the crystal [13]. The conversion efficiency improves by a factor of at least ten at liquid nitrogen temperatures. The sensitive detection of terahertz wave is also essential for many applications. A number of applications require sensitive, spectroscopic and wideband terahertz wave detector such as wireless communication and non-destructive inspection. In the future, we will endeavor to produce high sensitivity and wide bands for applied researches. This system could be powerful tools not only for solving fundamental physics but also real world problems such as manipulation or alteration of atoms, molecules, chemical materials, proteins, cells, chemical reactions, and biological processes, and real-time spectroscopic imaging, remote sensing, 3D-fabrication.

Acknowledgements

The authors would like to thank Dr. Minamide, Dr. Nawata, Dr. Taira. Prof. Ito of RIKEN, Prof. Kawase of Nagoya University. This work was partially supported by JSPS KAKENHI Grant Number 18H01908, and Tohoku University - NICT Matching.

Author details

Shin'ichiro Hayashi* and Norihiko Sekine
National Institute of Information and Communications Technology, Tokyo, Japan

*Address all correspondence to: hayashi@nict.go.jp

References

[1] M. Tonouchi, "Cutting-edge terahertz technology," Nat. Photonics, vol. 1, pp. 97-105 (2007).

[2] J. M. Chamberlain, "Where optics meets electronics: recent progress in decreasing the terahertz gap," Philos. Trans. R. Soc. London A, vol. 362, pp. 199-213 (2004).

[3] M. Sherwin, "Terahertz power," Nature, vol. 420, pp. 131-132 (2002).

[4] E. Linfield, "Terahertz applications - A source of fresh hope," Nat. Photonics, vol. 1, pp. 257-258 (2007).

[5] C. H. Henry and C. G. B. Garrett, "Theory of a parametric gain near a lattice resonance," Phys. Rev., vol. 171, pp. 1058-1064, (1968).

[6] M. A. Piestrup, R. N. Fleming, and R. H. Pantell, "Continuously tunable submillimeter wave source," Appl. Phys. Lett., vol. 26, no. 8, pp. 418-421, (1975).

[7] K. Kawase, M. Sato, T. Taniuchi, and H. Ito, "Coherent tunable THz wave generation from $LiNbO_3$ with monolithic grating coupler," Appl. Phys. Lett., vol. 68, no. 18, pp. 2483-2485, (1996).

[8] S. Hayashi et al., "Ultrabright continuously tunable terahertz-wave generation at room temperature," *Sci. Rep.*, vol. 4, Art. No. 5045, (2014).

[9] R. Guo, S. Ohno, H. Minamide, T. Ikari, and H. Ito, "Highly sensitive coherent detection of terahertz waves at room temperature using a parametric process," Appl. Phys. Lett., **93**, 021106 (2008).

[10] Kosuke Murate, Shin'ichiro Hayashi, Kodo Kawase, "Multi-wavelength terahertz-wave parametric generator for one-pulse spectroscopy," *Applied Physics Express*, vol. 9, pp. 082401 (2016).

[11] J. Shikata, K. Kawase, T. Taniuchi, and H. Ito, "Fourier transform spectrometer with a terahertz-wave parametric generator," Jpn. J. Appl. Phys., vol. 41 (Part 1, No. 1), pp. 134-138 (2002).

[12] L. Consolino, S. Bartalini, and P. De Natale, "Terahertz Frequency Metrology for Spectroscopic Applications: a Review," J. Infrared Milli. Terahz. Waves, vol. 38, pp. 1289-1315 (2017).

[13] J. Shikata, M. Sato, T. Taniuchi, H. Ito, and K. Kawase, "Enhancement of terahertz wave output from $LiNbO_3$ optical parametric oscillator by cryogenic cooling," Opt. Lett., vol. 24, pp. 202-204 (1999).

Chapter 8

A Novel Approach for Room-Temperature Intersubband Transition in GaN HEMT for Terahertz Applications

Rakesh Kaneriya, Gunjan Rastogi, Palash Basu, Rajesh Upadhyay and Apurba Bhattacharya

Abstract

Terahertz (THz) technology has attracted tremendous attention recently due to its promising applications in various domains such as medical, biological, industrial imaging, broadband, safety, communication, radar, space science, and so on. Due to non-availability of powerful sources and highly sensitive and efficient detectors, the so-called THz gap remains largely unfilled. Despite seamless efforts from electronics and photonics technology researchers, the desired level of technology development to fill the THz gap still remains a challenge. GaN-based HEMT structures have been investigated as potential THz sources and detectors by a number of researchers. This chapter presents a very new and versatile mechanism for electrical tuning of intersubband transitions (ISBT) GaN high electron mobility transition (HEMT) devices. ISBT phenomena are usually demonstrated in photonic devices like a quantum cascade laser (QCL). Here we explore ISBT in an electronic GaN HEMT device. Conventional photonic devices like a QCL are operated at cryogenic temperature to minimize thermal effect. Tuning the conduction band through external gate bias is an advantage of an HEMT device for room temperature (RT) THz applications. This chapter demonstrates the theoretical and experimental novel ISBT phenomenon in GaN HEMT is for potential ambient applications in the THz range.

Keywords: GaN HEMT, terahertz device, Intersubband transition, Plasmonic Metamaterials, quantum well

1. Introduction

Terahertz (THz) radiation is a small portion of the electromagnetic spectrum lying between the microwave and infrared regions. There is no precise range defining THz band, but it is most often –considered as frequencies in the range of 0.3–3.0 THz. Although sometimes it refers to 0.1–10 THz as well. The THz spectral range has drawn tremendous attention recently due to its promising applications in various domains. For example, in the field of biomedicine, THz radiation has been explored to detect various biomaterials like nucleic acids, proteins, cells and tissue applications [1, 2]. In the field of medical applications, the THz system has been demonstrated as a highly effective technique in cancer imaging, particularly for skin cancer [3, 4]. A portable real-time THz imaging system could be used to assist early detection of diseases during routine health checkups. Since many non-metallic, non-polar materials are transparent to THz radiation, scanning of humans is feasible with no health hazards. Due to this, THz radiation is widely used for security and public safety applications. THz radiation can detect concealed weapons, explosives (e.g., C-4, HMX, RDX and TNT), illicit drugs (e.g., methamphetamine and heroin), and more [1]. THz imaging has become a valuable characterization tool for non-destructive testing, process control and quality inspection for inspection of silicon solar cells, nanocomposites, polymer films and dielectric films [1, 5, 6]. Space- and ground-based THz instruments have been explored significantly in the field of astronomy. For example, the THz system is extensively used to study the origin of the universe, formation of stars and galaxies, composition of planets and planetary atmospheres, the climate and environmental balance of our planet Earth, and more [7, 8].

Despite these tremendous potential applications, the so-called THz Gap is not fulfilled to the required level due to technology requirements of high-power sources and efficient and sensitive detectors in the THz range. Semiconductor devices and circuits like transistor and frequency multipliers work well towards the low end of THz frequency, but their power level drops off precipitously as the frequency increases. These devices can be operated up to ~1 THz with very low power. Conversely, semiconductor photonic devices like lasers can be utilized in the high-frequency THz range. Again, lasers are limited due to the non-availability of lower bandgap semiconductor materials towards low-frequency THz. The THz quantum cascade lasers (QCLs) showed promising results to fill this THz gap from 1 to 10 THz. However, QCLs required bulky cooling requirements, and reported maximum operating temperature is in the range of 150–200 K, which is too low for general applications [9, 10]. The demand for a compact, efficient and high-speed THz detector and source operating at room temperature has increased drastically. The non-availability of a room-temperature THz source and detector is a prime limitation of the modern THz system.

In this chapter, novel theoretical models and experimental techniques for the intersubband transitions (ISBT) phenomenon are illustrated for ambient THz applications. Section 2 covers the theoretical models and simulations based upon plasmonic metamaterials-assisted ISBT and describes the GaN HEMT response towards the THz spectrum. Section 3 covers the fabrication and measurement of a GaN HEMT device. Section 4 is investigates ambient temperature ISBT in a GaN HEMT device.

2. Theoretical modeling and simulation strategy

In this section, we present our theoretical model based on ISBT, metamaterial and plasmonic phenomena for GaN HEMT THz applications [11–13]. We proposed a combined plasmonic and metamaterial-driven ISBT phenomenon as one of the possible modes that can extend GaN HEMT operating frequency well beyond its present cut-off frequency to the THz band. ISBT is the prime mechanism to explore as a potential mechanism for THz operation, while metamaterial and plasmonic effects improve the strength of ISBT in a GaN HEMT structure. Theoretical modeling started with the role of polarization in a wurtzite semiconductor followed by the self-consistent solution of Schrodinger and Poisson equations; k.p model and Fermi Golden rule are used to compute ISBT in the GaN HEMT structure. The size and geometry of an HEMT device act as THz metamaterial (this concept is explored in-depth in Section 2.4) and it couples THz radiation to two-dimensional electron gas (2DEG) inside a triangular quantum well. Further, very small gate lengths in the range of 100–250 nm are selected for high-frequency operation of HEMT. This fine nanometric-sized gate structure of HEMT excites surface plasmon waves at the interface between the gate metallic contact and 2DEG channel in the GaN heterostructure (this concept is explored in-depth in Section 2.5). These combinations (i.e., plasmonic metamaterial-assisted ISBT) govern the THz response of the GaN HEMT device.

2.1 GaN heterostructure and device description

GaN heterostructure is generally grown on sapphire or silicon carbide (SiC) substrate. **Figure 1(a)** shows the most widely used GaN heterostructure, which consists of a 60-nm AlN nucleation layer, 2-μm thick undoped GaN layer, 1-nm AlN spacer layer, 20 nm-undoped $Al_{0.3}Ga_{0.7}N$ barrier layer, and 3-nm Si_3N_4 passivation layer. Introducing a thin 1-nm AlN interlayer between AlGaN and GaN plays a crucial role. Better carrier confinement, reduced alloy scattering and enhanced conductivity are achieved by inserting a thin AlN layer [14, 15]. The cross-sectional view of the simulated GaN HEMT device by Silvaco TCAD is shown in **Figure 1(b)**. Computation mesh to simulate the device structure is shown in **Figure 1(c)**. In the regions beneath the gate, at the edges of the source and drain contacts and at the AlGaN/AlN/GaN interface, fine meshing is done to achieve the convergence and accuracy of the calculations. The spacing between different electrodes, namely, source to gate, gate to drain and source to drain are set to 0.9, 2.0 and 3.0 μm, respectively. Gate length is kept as 100 nm. To obtain lower gate resistance, gate geometry is selected as T-gate in simulation as well as in fabrication.

Generally, high-power RF GaN HEMT is fabricated in a multi-finger configuration. Two ground-source-ground (GSG) configurations are shown in **Figure 2**: 2 × 150 and 8 × 150. To measure the RF performance of the device GSG configuration is widely used for HEMT fabrication. The 2 × 150 configuration contains two gate fingers with 150-micron unit gate width of the device. Similarly, the 8 × 150 configuration contains eight gate fingers with 150-micron unit gate width. To expand the device length for high-power applications, a greater number of gate fingers are used. For example, if the power handling capability of the fabricated GaN HEMT is 5 W/mm, the 2 × 150 = 0.3-mm device can be used for 1.5 W RF power. Similarly, the 8 × 150 = 1.2-mm device can be used for 6.0 W RF power. Before further discussion on plasmonic metamaterial-assisted ISBT, the following section refreshes some fundamentals about polarization in III-N (nitride) semiconductors.

2.2 Polarization in GaN heterostructure

The nitride semiconductor materials exhibit inherent polarization properties. Having the large ionicity of the nitride bond (Ga–N, Al–N, In–N, etc.), it possesses a

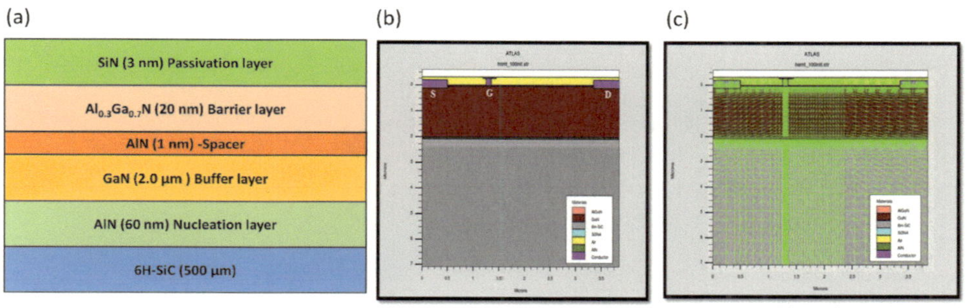

Figure 1.
(a) GaN Heterostructure, (b) HEMT cross-sectional view and (c) HEMT mesh structure (reprinted with permission from Ref. [12]).

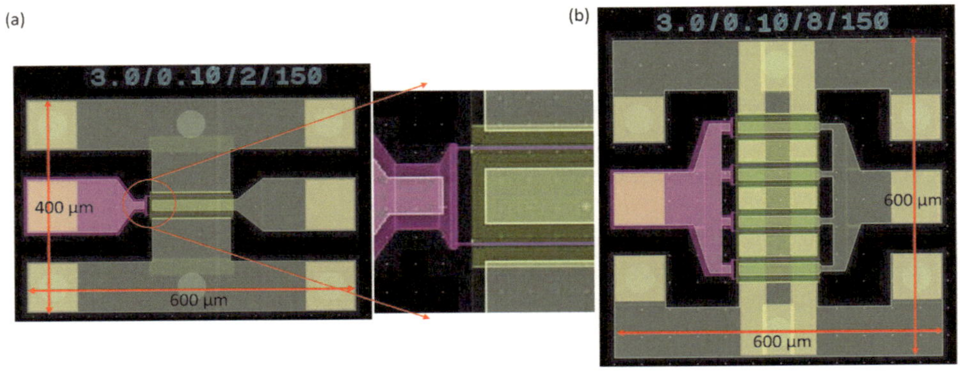

Figure 2.
Typical configurations of GaN HEMT (a) 2 gate fingers with 150 micron gate width (2 × 150) (zoom image for gate fingers visualization and (b) 8 gate fingers with 150 micron gate width (8 × 150). Representation 3.0/0.10/8/150 indicates source to drain distance 3.0 µm, 0.1 µm (100 nm) gate length, 8 fingers device with 150 µm gate width.

piezoelectric polarization (P_{PE}) component, while the absence of the center of inversion symmetry and uniaxial nature of the crystal structure produces spontaneous polarization (P_{SP}). Total polarization (P_T) in the nitride semiconductor heterostructure is a combination of spontaneous polarization (P_{SP}) and piezoelectric polarization (P_{PE}), as shown in Eq. (1).

$$P_T = P_{SP} + P_{PE} \tag{1}$$

Furthermore, the strain-induced effect at the interface between two nitride semiconductors enhances piezoelectric polarization in the heterostructure. Piezoelectric polarization of the crystal is generally defined in terms of strain (ε) and stress (σ) components. Stress and strain are correlated in a crystal by elastic coefficient $\varepsilon_{ij} = C_{ij}\,\sigma_{ij}$. The piezoelectric polarization in heterostructure grown along the z-axis (0001) is given by,

$$P_{PE} = E_{33}\varepsilon_z + E_{31}\left(\varepsilon_x + \varepsilon_y\right) \tag{2}$$

where E_{33}, and E_{31} are piezoelectric coefficients, and ε_x, ε_y and ε_z are strain in x, y and z-directions, respectively. The crystal edge length and height are represented as a_0 and c_0 respectively in a hexagonal crystal lattice. The strain along the x, y and z-axis is given by (in-plane strain along x-axis and y-axis are assumed to be isotropic),

$$\varepsilon_z = \frac{c - c_0}{c_0}, \varepsilon_x = \varepsilon_y = \frac{a - a_0}{a_0} \tag{3}$$

where a_0 and c_0 are the equilibrium or unstrained values of lattice constants and a and c are the strain lattice constant due to growth of heterostructure. For hexagonal lattice crystal, the strain components along ε_z and ε_x are related with elastic coefficients as per the following equation,

$$\varepsilon_z = -2\frac{c_{13}}{c_{33}}\varepsilon_x, \quad \frac{c - c_0}{c_0} = -2\frac{c_{13}}{c_{33}}\frac{a - a_0}{a_0} \tag{4}$$

where C_{13} and C_{33} are the elastic constants. Substituting Eqs. (3) and (4) in Eq. (2),

$$P_{PE} = 2\frac{a - a_0}{a_0}\left(E_{31} - E_{33}\frac{c_{13}}{c_{33}}\right) \tag{5}$$

The macroscopic piezoelectric polarization is defined by variations of the lattice constants a and c. The microscopic piezoelectric polarization is expressed in terms of an internal parameter u, defined as the anion–cation bond length along the z-axis (0001) [16]. Substituting elastic constant values for AlN and GaN in Eq. (5), one gets piezoelectric polarization of AlN greater than GaN. Spontaneous polarization closely depends upon crystal structure c/a ratio. The ideal c/a ratio in the hexagonal, closed-pack crystal structure is 1.633. The spontaneous polarization is found to be greater in actual crystal structures as the c/a ratio is different from its ideal value [16]. This nonideality of c/a ratio in AlN is also greater than GaN, which leads the greater spontaneous polarization. The spontaneous and piezoelectric polarization for alloy (i.e., AlGaN) is obtained by linear interpolation of the binary constituents (Vegard's law). In summary, the spontaneous and piezoelectric polarizations for AlGaN over the whole range of compositions are larger than that of a GaN buffer layer.

The polarization-induced charge density and sheet density in the heterostructure is given by,

$$\rho = -\nabla * P \tag{6}$$

$$\sigma = P_{T(Layer1)} - P_{T(Layer2)}$$

$$\sigma = \left[P_{SP(Layer1)} + P_{PE(Layer1)}\right] - \left[P_{SP(Layer2)} + P_{PE(Layer2)}\right] \tag{7}$$

The polarization-induced charge density and sheet density for the case of AlGaN/GaN heterostructure is given by,

$$\sigma = \left[P_{SP(GaN)} + P_{PE(GaN)}\right] - \left[P_{SP(AlGaN)} + P_{PE(AlGaN)}\right] \tag{8}$$

Extracted 2DEG concentration, purely due to polarization effects, is the order of $\sim 10^{13}$ cm^{-2} for nitride heterostructures. Unlike GaAs MODFET heterostructures, no doping is required in nitride heterostructures to generate 2DEG concentration, which is a great advantage of these structures.

The basic equations of physical processes are solved for every grid point in the simulation. These equations include Poisson's equation, continuity equations and transport equations, derived from Maxwell's equations [17]. The computation of 2DEG properties due to spontaneous and piezoelectric polarization effects is performed using a polarization model [18, 19]. An induced, strong polarization field is introduced to calculate band diagrams. To increase the reliability of simulation, measurement-based ohmic contact resistance and Schottky barrier height data were incorporated in the simulation to define source, drain and gate contacts. A low field mobility model is used to account for the temperature-dependent drift of electrons and holes separately [20]. The Shockley–Read–Hall recombination model is used to estimate the statistics of holes and electrons as well as their recombination rate. The traps/defects in the heterostructure play a crucial role in the performance of GaN devices. Accordingly, we also introduced interface traps energy level and density in the modeling. Output results were extracted by solving the basic equations for every grid point with the different biasing conditions. The variation of the drain current with respect to applied drain (Vd) and gate (Vg) biasing voltage is plotted in **Figure 3**. The simulated output characteristics (Id-Vd) and transfer characteristics (Id-Vg) are shown in **Figure 3(a)** and **(b)**, respectively. The extracted transconductance is >350 mS/mm as shown in **Figure 3(c)**. The extracted capacitance-gate

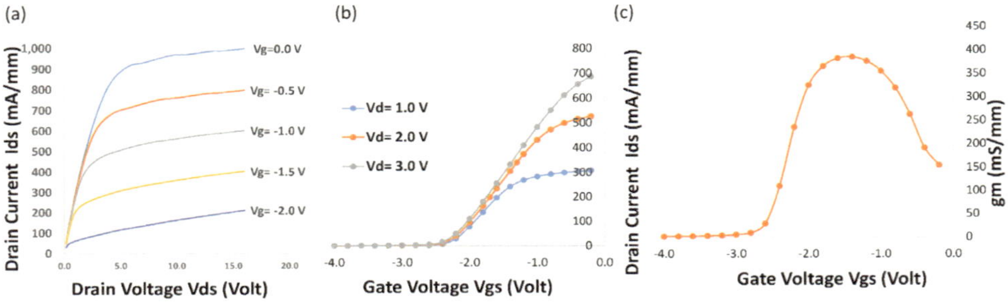

Figure 3.
(a) Output characteristics (id-Vd), (b) transfer characteristics (id-Vg) and (c) Transconductance of the simulated device (reprinted with permission from Ref. [11]).

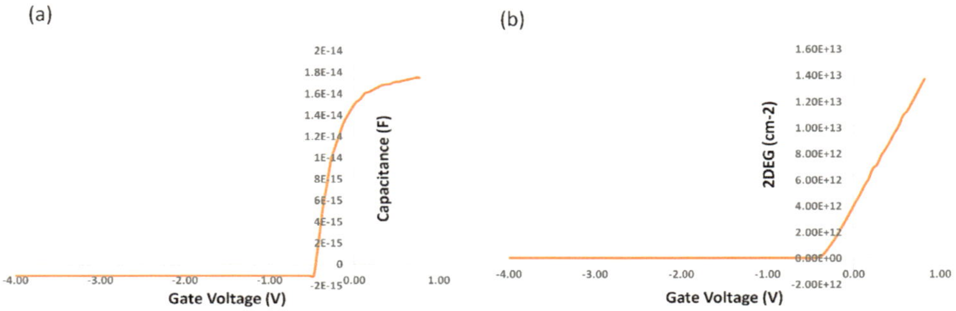

Figure 4.
Extracted (a) CV profile and (b) 2DEG profile with applied gate voltage (reprinted with permission from Ref. [12]).

voltage (Vg) and 2DEG density with applied gate bias are depicted in **Figure 4(a)** and **(b)**, respectively.

The current gain cutoff frequency (f_t) and maximum frequency of oscillations (f_{max}) are the two most pertinent parameters for high-speed device application. f_t and f_{max} are extracted from small signal RF simulation. Current gain (h_{21}) and maximum available power gain (G_a) are simulated at bias conditions Vds = 7 V and Vgs = −1.5 V and plotted with respect to frequency in **Figure 5**. A summary of simulated DC and RF device parameters is given in **Table 1** that closely matches the corresponding process design kit (PDK) datasheet of renowned international GaN foundries.

The cutoff frequency of field effect transistor (FET) including HEMT is defined by $f_T = \upsilon/2\pi L = 1/2\pi\tau$, where υ is carrier velocity, L is gate length and τ is electron transit time under the gate. For very small gate length of 30 nm, up to 300–350 GHz cut-off frequency operation has been demonstrated [21]. However, beyond conventional transit-time limitations, these FET devices can be operated at much higher frequencies up to THz. Dyakonov–Shur proposed the plasma wave theory to describe THz behavior in FET devices [22, 23]. The basis of plasma wave theory is the instability of 2DEG, which has a resonant response to incident electromagnetic radiation in short-channel FET. The size and shape of the FET channel are used to govern the resonant response of plasma frequency to electromagnetic radiation. The tuning of plasmon frequency by external biasing has been used for detectors, mixers and multipliers [23]. There are several reports available in which the plasma wave theory is used to describe the THz behavior of devices. The plasma wave theory concept has been widely demonstrated in conventional semiconductors like Si [24], GaAs [22], GaN [25], and InP [26] as well as in new two-dimensional

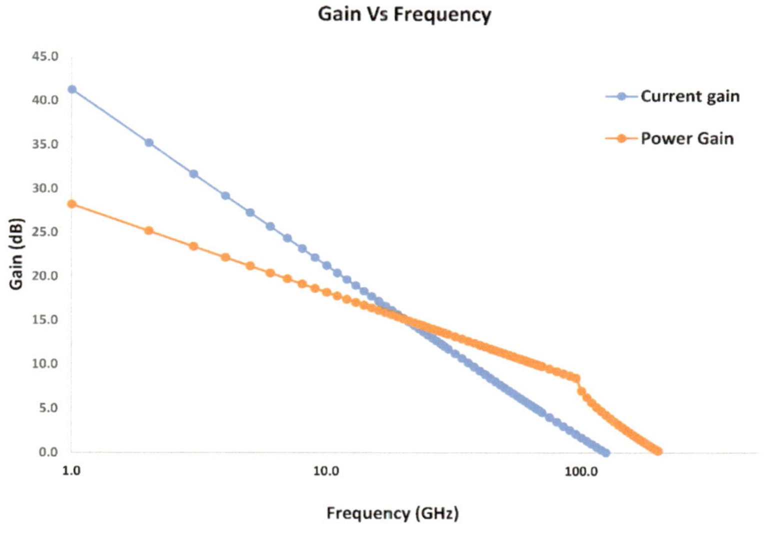

Figure 5.
Current gain and power gain of the simulated device (reprinted with permission from Ref. [11]).

	Simulated result
Idss (A/mm)	0.995
Vknee (V)	5
Ron (ohm*mm)	3.5
Vth (V)	−3.0
gm (mS/mm)	384
ft (GHz)	110
fmax (GHz)	180

Table 1.
Summary of extracted DC and RF device parameters.

materials systems like graphene [27], black phosphorus [28], and more. There are a few papers available in which a working GaN HEMT mechanism is explained through plasma wave theory [25, 29–32]. Based on our simulation and experimental investigations, we have demonstrated for the first time that an ISBT mechanism, in addition to established plasma wave theory, can describe the THz behavior of a GaN HEMT device. ISBT theory is based on carrier transitions within a conduction band, which is entirely different from plasma wave theory. The fundamentals of this theory are explained below.

2.3 Quantum confinement in GaN heterostructure

The interaction between photons and electrons in the semiconductor can be expressed by the Hamiltonian,

$$H = \frac{1}{2m_0}(\mathrm{p} - \mathrm{eA})^2 + V(r) \tag{9}$$

where m_o is the free electron mass, V(r) is the periodic crystal potential (in the present case it is the triangular potential function given by, V(z) = eFz), e is charge of electron, Fz is electric field, and A is vector potential of applied electromagnetic field. Hamilton can be expanded into,

$$H = \frac{p^2}{2m_0} + V(r) - \frac{e}{2m_0}(\mathrm{p.A} + \mathrm{A.p}) + \frac{e^2A^2}{2m_0} \tag{10}$$

$$H \approx H_0 + H'$$

Here H_0 is unperturbed Hamiltonian and H′ is perturbed Hamiltonian due to the interaction of the electromagnetic wave.

Consideration of the strain effects for extraction of effective-mass Hamiltonian is of prime importance for wurtzite semiconductors. This Hamiltonian is used to derive the electronic band structures of bulk and quantum-well wurtzite semiconductors. Kane's model is applied to derive the band-edge energies and the optical momentum-matrix elements for strained wurtzite semiconductors. We then derive the effective-mass Hamiltonian by using the k.p perturbation theory. The developed k.p model is applied to our heterostructures structures, especially quantum well via the envelope function approximation (EFA) method [33, 34]. An envelope function model is derived for electrons in a semiconductor heterostructure. The materials-dependent Hamiltonian extraction by EFA method is most suitable for abrupt semiconductor junction [35]. The finite element method [36] is used to solve the coupled multi-band Schrödinger Poison's equation [37] numerically.

Under triangular quantum well, the solution of the wave function is given by [38, 39],

$$\Psi(z) = Ai\left[\frac{2m_z{}^{*}eF_z}{\hbar^2}\left(\mathrm{z} - \frac{E_i}{eF_z}\right)\right] \tag{11}$$

where $m_z{}^*$ is the effective mass of electron in the GaN, Fz is the electric field in the z-direction, Ei is the eigenvalues of energy with i = 0,1,2, … . for the ground state, 1st excited state and so on. Airy (Ai) function is given by

$$Ai(z) = \frac{1}{\pi}\int_0^{\infty} \cos\left(\frac{t^3}{3} + zt\right) dt$$

The eigen value is given by [37, 38],

$$E_i \approx \left(\frac{\hbar^2}{2m_z{}^{*}}\right)^{1/3}\left[\frac{3\pi eF_z}{2}\left(i + \frac{3}{4}\right)\right]^{2/3} \tag{12}$$

When an incident THz radiation illuminates the GaN HEMT, electrons may absorb the photon energy and jump to a higher energy subband. Carriers below Fermi energy levels were collected by drain electrode when we applied voltage between source and drain. Using Fermi's golden rule for the transition from i state to j state, we can calculate the absorption coefficient by [34, 40],

$$W_{ij} = \frac{2\pi}{\hbar} < \Psi_i|H'|\Psi_j >^2 \delta(E_j - E_i - \hbar\omega)$$

where H′ is interaction Hamiltonian as per Eq. (10).

By applying the dipole approximation, we obtain [34, 40],

$$W_{ij} = \frac{2\pi}{\hbar}\frac{e^2E_0{}^2}{4m_z{}^2\omega^2} < i|e.p|j >^2 \delta(E_j - E_i - \hbar\omega) \tag{13}$$

The matrix element in the above equation can be expanded in terms of interband and ISBT as follows,

$$<i|e.p|f> = e.<u_v|p|u_{v'}> <f_n|f_{n'}> + e.<u_v|u_{v'}> <f_n|p|f_{n'}>$$

Applying the envelope function matrix element in the z-direction can be written as

$$<i|p_z|j> = \int \Psi_i^*(z) p_z \Psi_j(z) dz$$

The dimensionless optical field strength between the two-energy state is given by [33, 38],

$$f_{ij} = \frac{2}{m_z{}^* \hbar\omega_{ij}} <i|p_z|j>^2 = \frac{2m_0{}^* \omega_{ij}}{\hbar} <i|z|j>^2 \tag{14}$$

where $\omega_{ij} = (E_j - E_i)/2$
$<i/z/j>$ can be expressed as,

$$<i|z|j> = z_{ji} = \frac{2L}{(t_i - t_j)^2} \tag{15}$$

with, ti and L are electric length expressed as,

$$t_i = \left[\frac{3\Pi}{2}\left(i + \frac{3}{4}\right)\right]^{2/3} \tag{16}$$

$$L = \left(\frac{\hbar^2}{2m_z{}^* eF_z}\right)^3 \tag{17}$$

By substituting Eqs. (15)–(17) in Eq. (14) we get,

$$f_{ij} = \frac{8.32}{(t_i - t_j)^4}\left(\frac{\Pi}{2}\right)^{2/3}\left[\left(j + \frac{3}{4}\right)^{2/3} - \left(i + \frac{3}{4}\right)^{2/3}\right] \tag{18}$$

By substituting i = 0 and j = 1,2,3, ... oscillation strength for transition can be calculated as f_{01} = 0.73, f_{02} = 0.12, f_{03} = 0.045, and so on. The oscillator strength of all the transitions is sum up to 1. Calculated transition indicates that the probability for higher-level transitions is very weak.

The gradual pinning of Fermi level inside the quantum well is possible by increasing gate voltage. When gate voltage is sufficiently negative (0 > Vt > Vg), the conduction band is above the Fermi level. In this case, the channel is completely depleted of 2DEG. When the gate voltage is greater than the threshold voltage (Vg > Vt), charges start filling the channel. As the gate voltage increases, the Fermi level gradually pins inside the quantum well and 2DEG carriers are filled among allowed subbands in the channel. When gate voltage is sufficiently higher (Vg > 0 > Vt), the carrier occupies all allowed subband below the Fermi energy level. For this case, total 2DEG charges are distributed in the allowed energy subband and take participation in channel conduction. The triangular quantum-well conduction band energy profile for GaN HEMT with different gate biasing conditions is shown in **Figure 6(a)**. Fermi energy level pinning inside the subbands of triangular quantum well with different applied gate biasing is shown in

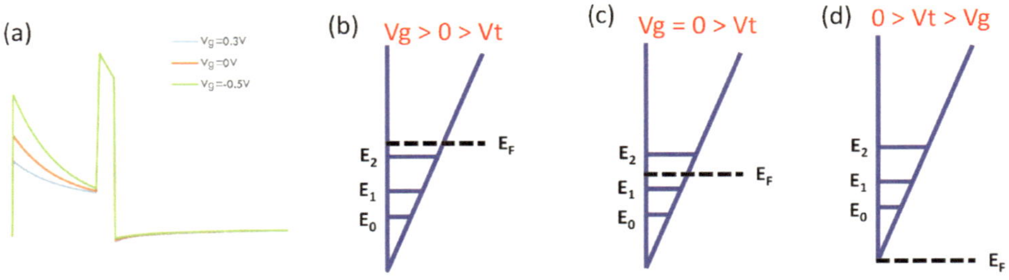

Figure 6.
Conduction band energy profile with different applied gate voltage and Fermi energy level with filled subband inside triangular quantum well with different applied gate biasing (b) Vg > 0 > Vt, (c) Vg = 0 > Vt and (d) 0 > Vt > Vg (reprinted with permission from Ref. [12]).

Figure 6(b)–(d). The spacing and charge filing inside the subband strongly depends upon gate-biasing voltage. In other words, the gate biasing-assisted tuning of intersubband resonance (ISR) frequency is possible in the HEMT structure.

In the simulation, we extracted up to four ISB energy levels inside the triangular quantum well. The ISR frequency as a function of applied gate-biasing field is calculated using Eq. (12) and by solving self-consistency Schrodinger–Poison solver for different gate-biasing voltage. The same are shown in **Figure 7(a)** and **(b)**. The ISB tuning is one order higher in asymmetric triangular well potential as compared to the conventional square well potential. Moreover, 2DEG carrier concentration inside the GaN HEMT channel also depends upon Al composition and AlGaN barrier layer thickness. **Figure 8(a)** and **(b)** show the simulated 2DEG carrier concentration variation with AlGaN thickness and Al composition, respectively. It clearly indicates that increment in barrier layer thickness, and Al composition enhances the 2DEG density inside the channel. It further implies that manipulation in ISR is possible in GaN HEMT devices based on variation in 2DEG density, which provide tuning in the THz region.

2.4 Metamaterial-embedded ISBT

The combination of ISBT in semiconductor quantum wells with metamaterials shows great potential in the THz region [41–47]. There are large numbers of metamaterial structures that have been employed and demonstrated enhanced performance in the THz region. In the present modeling work, we report that the standard GSG device geometry of HEMT itself acts as a metamaterial structure. The enhancement of THz interaction with 2DEG inside the triangular quantum well is

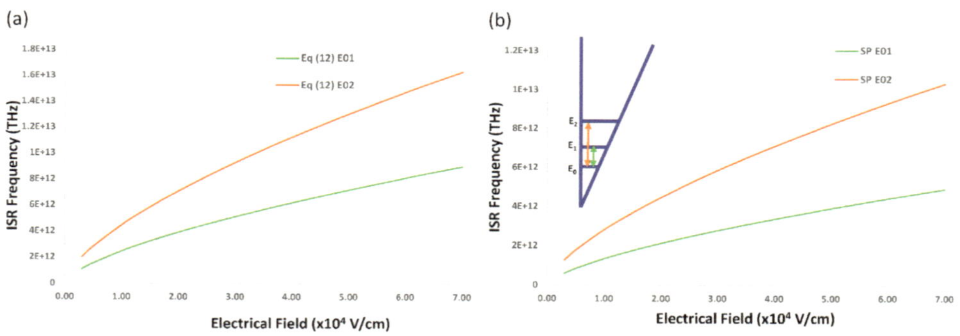

Figure 7.
Intersubband resonance frequency as a function of the applied field (a) calculated using Eq. (12), (b) by solving self-consistency Schrodinger-Poisson solver. (reprinted with permission from Ref. [12]).

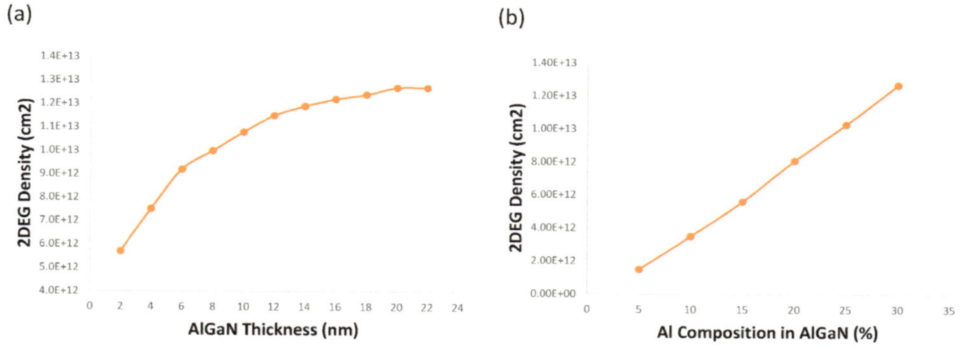

Figure 8.
2DEG carrier concentration of AlGaN/AlN/GaN heterostructure for (a) different AlGaN thickness, (b) different Al composition in AlGaN layer (reprinted with permission from Ref. [12]).

reported for GaN HEMT. The resonance mode in metamaterial structure is dynamically manipulating the carrier distribution inside the quantum well.

For the metamaterial modeling work, the GaN heterostructure and device geometry are kept identical, as shown in **Figure 1**. A finite difference frequency domain CST Microwave Studio simulator has been used to simulate the entire device configuration, which acts as THz metamaterial. Standard GSG configuration along with 50- to 150-micron gate width has been used for 3D electromagnetic modeling as shown in **Figure 9(a)–(c)**. Very fine localized tetrahedral sub-meshing has been used in the active source to drain region to enhance the accuracy of calculations as shown in **Figure 9(d)**. THz radiation (0.3–3 THz) is illuminated on the entire GSG device configuration, which includes the active GaN HEMT region as well. **Table 2** shows the dimensions used in 3D EM simulation work. Three different geometries, 2×50, 2×100 and 2×150, have been used in the present study. The E-field of the incident THz plane wave is kept at 1 V/m for all three devices.

The wavelength corresponding to the entire THz spectrum (0.3 to 10 THz) is about 30–1000 micron. If the device dimension is of the order of incident radiation, it acts as an antenna. Antenna size and shape largely determine the frequency it can handle. Antenna-coupled THz source and detector show a potential advantage in the performance of devices for the THz region [48–52]. The dimensions of the devices as listed in **Table 2** are of the order of illuminated THz radiation wavelength. These devices act as antennas, which leads to convergence of incident radiations towards the active channel region. The resultant electric field intensity inside the active channel region between source and drain is greatly enhanced. The enhancement of the field due to illumination strongly depends on the frequency of incident radiation and device dimension. For example, the electric field intensity distribution for 0.4 THz incident radiation is shown in **Figure 10** (a)–(c) for three different GaN HEMT devices. Each device structure has a unique resonance response towards incident THz radiation. Similarly, the resonance response of a 2×100 GaN HEMT device towards incident THz radiations, namely 0.3, 0.7 and 1.75 THz, is shown in **Figure 10(d)–(f)**. Moreover, the illumination-dependent enhancement of the field is not distributed uniformly throughout the channel. The highest field distribution found at the center of a 2×100 device for 1.75 THz illuminations is shown in **Figure 10(f)**. The summary for average field enhancement inside the channel region due to illumination is illustrated in **Figure 11**. The GaN HEMT device itself acts as metamaterial, which further influences the overall THz performance of the GaN HEMT device.

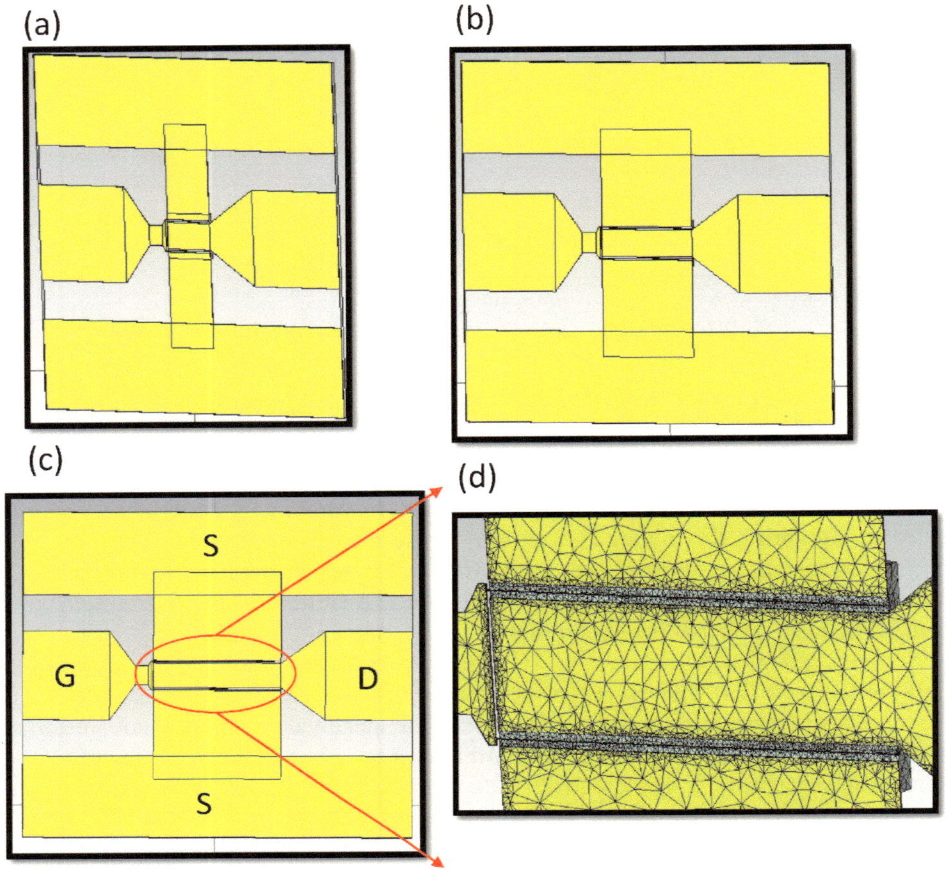

Figure 9.
Three different configuration (a) 2 × 50, (b) 2 × 100 and (c) 2 × 150 of GaN HEMT and (d) GaN HEMT meshing (reprinted with permission from URSI RCRS 2020, IEEE Xplore).

Device configuration	Distance between S and D (µm)	Gate width (µm)	Gate length (nm)	Total device dimension (µm x µm)
Device A (2 x 50)	3.0	50	100	350 x 400
Device B (2 x 100)	3.0	100	100	400 x 400
Device C (2 x 150)	3.0	150	100	450 x 400

Table 2.
Different device configurations used in simulations.

2.5 Plasmonic-assisted ISBT

A plasmonic nanostructure provides unique opportunities for manipulating electromagnetic waves in the THz range. Recently many novel plasmonic nanostructure-based devices such as photoconductor antennas [52, 53], detectors [31], and plasmonic photomixers [54], QCLs [55] showed significant improvement in device performance.

For plasmonic structure simulation, the finite element frequency domain COMSOL Multiphysics numerical method has been used to solve Maxwell's equation to predict electromagnetic interaction in each layer of the semiconductor

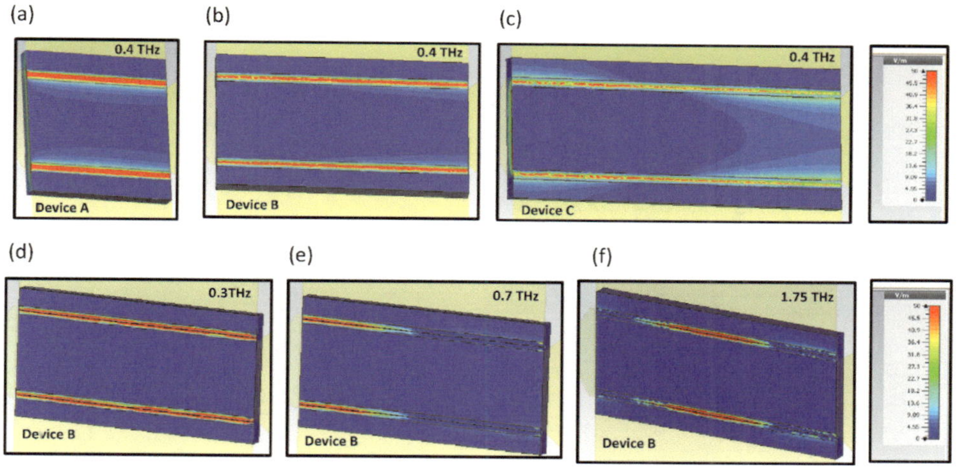

Figure 10.
Electrical field enhancement for 0.4 THz incident radiation on (a) 2 × 50, (b) 2 × 100 and (c) 2 × 150 of GaN HEMT devices. Electrical field enhancement for 2 × 100 GaN HEMT device at (d) 0.3°THz, (e) 0.7° THz and (f) 1.75°THz incident radiation (reprinted with permission from URSI RCRS 2020, IEEE Xplore).

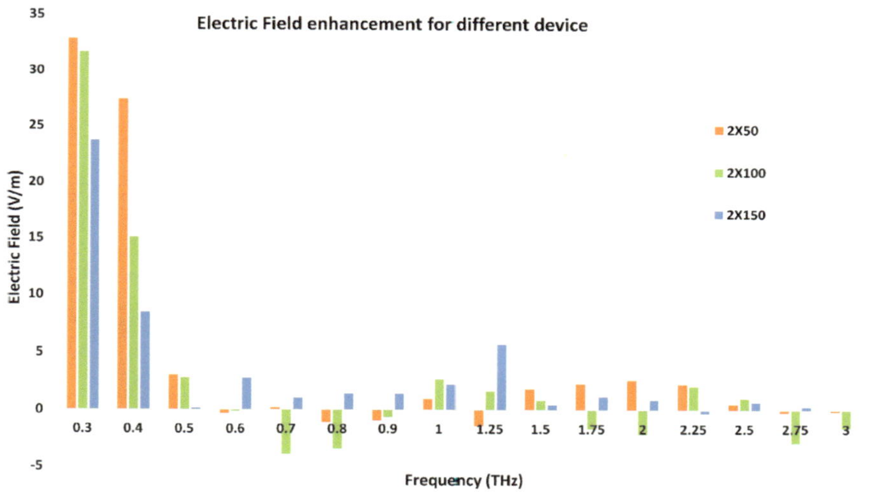

Figure 11.
Electrical field enhancement due to illumination of terahertz radiation on GaN HEMT devices.

heterostructure. Heterostructure stack, device geometry and device structure are kept identical as used in semiconductor modeling shown in **Figure 1**. We kept 1 V/m incident plane wave THz radiation from 0.3 to 3 THz to interact with the GaN heterostructure. The surface plasmon is generated at the interface between nanometric gate contact and heterostructure.

The field in the vicinity of the fine gate structure is drastically increased due to surface plasmon generation. Subsequently, the THz incident wave is coupled to 2DEG inside the channel. The concentration of the induced electric field is considerably enhanced in close proximity to the device gate contact electrodes. The induced electric field is approximately 5.5E+06 on the gate and 8.5E+06 V/m on the gate edge for 0.4 THz due to plasmonic structure as shown in **Figure 12**. As the incident frequency increases, the plasmonic-induced electric field also increases and saturates towards higher frequency as depicted in **Figure 13**. It was interesting to find that the plasmonic-enhanced field ($\sim 10^7$ V/m) is approximately one order

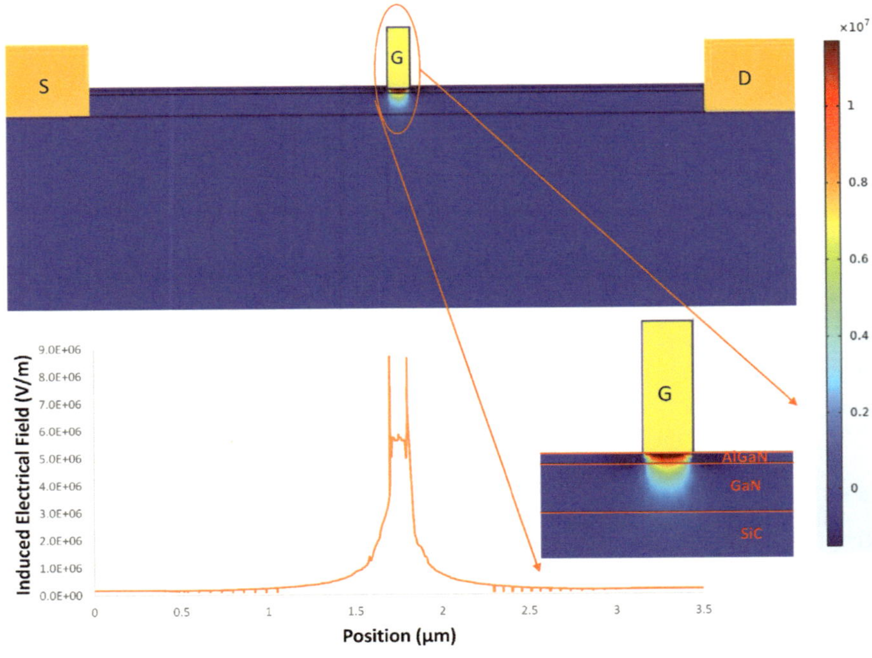

Figure 12.
Electrical field enhancement for 0.4°THz incident radiation on GaN HEMT device using a finite element method-based electromagnetic solver (COMSOL).

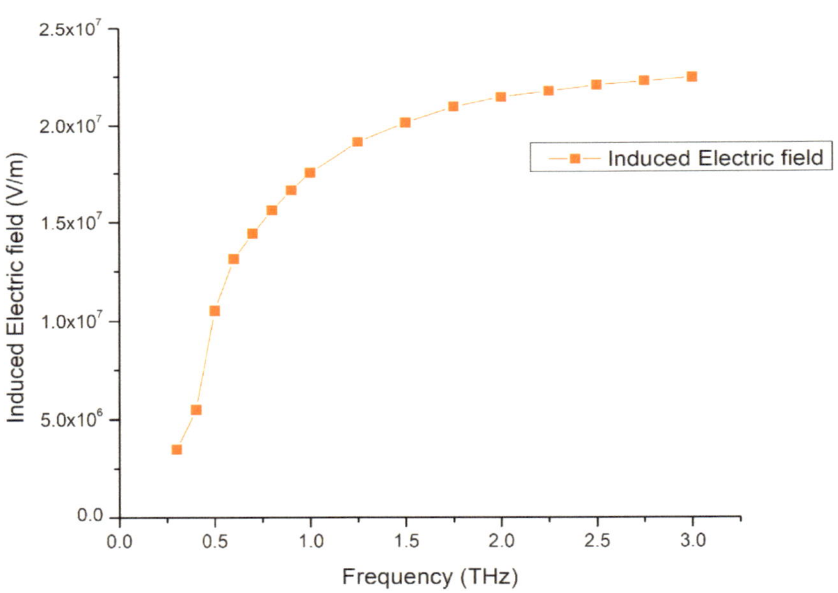

Figure 13.
Induced electrical field due to terahertz radiation illumination on GaN HEMT device using a finite element method-based electromagnetic solver (COMSOL).

greater than the externally applied bias field ($\sim 10^4$ V/cm = 10^6 V/m) at the gate (**Figure** 7).

The outcome of the entire simulation activities clearly demonstrates GaN HEMT device operation in the THz range beyond its cut-off frequency. It is also shown that overall performance of GaN HEMT is governed by aggregate effects of ISBT, plasmonic structure and metamaterial behavior.

3. Fabrication and measurement of GaN HEMT devices

In this section we report fabrication and measurement details for experimental investigation of room-temperature, photon-induced electrical tuning of ISBT in GaN HEMT, which extends the device operating frequency well beyond its present cut-off frequency [56]. For sample fabrication, an AlGaN/AlN/GaN-based heterostructure was grown by metalorganic chemical vapor deposition (MOCVD) on 6H polytype of silicon carbide (6H-SiC) wafer. The layer sequence, thickness and composition were kept identical as used in our modeling and simulation work (**Figure 1(a)**). A 60-nm aluminum nitride (AlN) nucleation layer was grown on (0001) the face of a semi-insulating silicon carbide (SI–SiC) wafer. In the MOCVD growth, unintentionally doped (UID) GaN buffer layer thickness was set to ~2 μm. On the top of the GaN buffer layer, a 1-nm AlN spacer layer followed by an undoped $Al_{0.3}Ga_{0.7}N$ barrier layer was grown to form a triangular quantum well of GaN HEMT. A small 3-nm Si_3N_4 passivation layer was kept as a top protective layer. For the purpose of characterization of this GaN heterostructure to assess quality and properties, highly precise standard semiconductor characterization tools like Hall measurement (nanomagnetic instruments), high-resolution XRD (Bruker D8 Discover), photoluminescence (PL) system (DongWoo Optron), and others were used. Room-temperature mobility and 2DEG carrier concentration were measured using the Hall measurement method. Composition, thickness and lattice constants for heterostructure materials were extracted using HR-XRD. The growth quality was evaluated using the PL method.

A standard fabrication process flow as shown in **Figure 14** was adopted for GaN HEMT device fabrication. Device-to-device isolation was performed by MESA

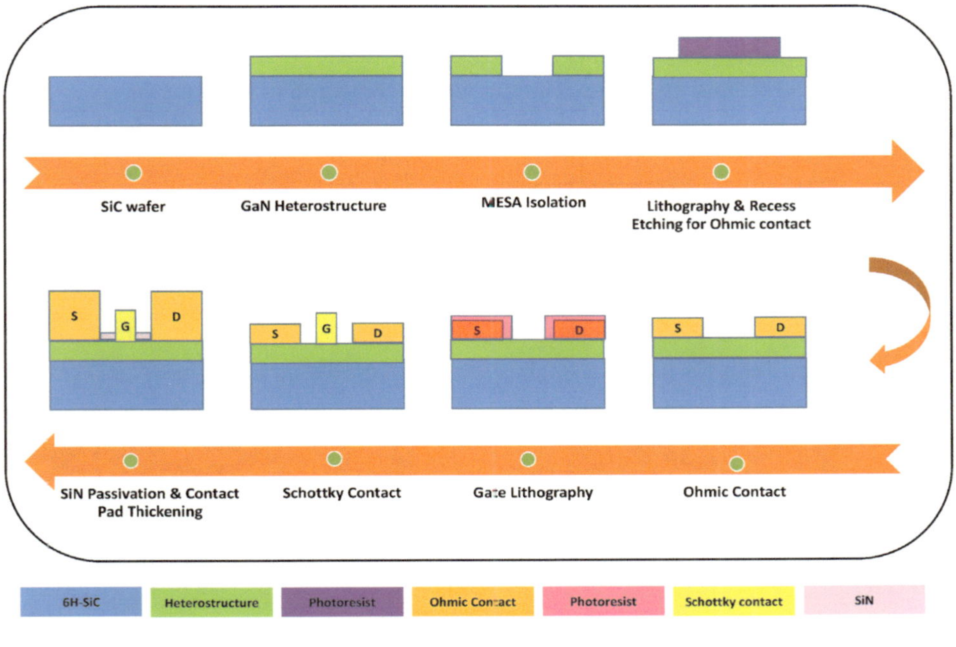

Figure 14.
GaN HEMT fabrication flow along with the cross-sectional view of the device.

etching using $BCl_3/Cl_2/Ar$ dry plasma. An inductively coupled plasma-reactive ion etching (ICP-RIE) system was used for HEMT device isolation. Source-drain spacing was kept at 3.0 micron and electron beam lithography was used for ohmic contact (source and drain) patterning. Recess etching of the barrier AlGaN layer was required to fabricate good ohmic contacts. An optimized recessed etching process was followed to etch ~10 nm of the AlGaN layer using $BCl_3/Cl_2/Ar$ plasma. During recess, $BCl_3/Cl_2/Ar$ flow rates were maintained at 20/10/10 SCCM with ICP power and RF power at 350 W and 60 W, respectively. Post recess, the sample was dipped in HCl:DI (1:10) for one minute to minimize the impact of oxidation on the surface. A Ti/Al/Ni/Au (20/210/55/45 nm) lift-off metallization scheme was selected and deposited by electron beam evaporation for ohmic contact. The sample was annealed at 870°C for 45 s under N2 atmosphere using rapid thermal annealing to form the ohmic contact [57]. The ohmic contact resistance was measured using a standard transmission line model (TLM) with the help of a semiconductor characterization system (Model Keithley 4200). Electron beam lithography was used to form a mushroom gate contact. A Ni/Au metallization scheme was selected for gate contact to achieve high Schottky barrier height. A Ni/Au (30/300 nm) stack was deposited using electron beam evaporation and lift-off technique followed by annealing at 450°C for 120 s under N2 atmosphere using rapid thermal annealing system [58]. To address the DC-RF dispersion issue, the Si_3N_4 passivation layer of 120-nm thickness was deposited using PECVD. Contact pad thickening was formed by 800-nm Ti/Au deposition to reduced resistive loss. The fabricated GaN HEMT wafer (1 square inch) having more than 300 GaN HEMT devices is shown in **Figure 15**. A sufficient number of variants are kept in fabrication in terms of device length and number of device fingers for wider statistical data. Device length varies from 50 to 300 micron (50, 100, 150, 200, 300), while number of fingers varies from 2 to 12 (2, 4, 6, 8, 10, 12). All devices have a 100-nm mushroom gate structure. The wafer also contains varieties of process control monitors (PCMs) for ohmic contact, Schottky contact, short, open and through the structure for RF measurement.

Current–voltage (IV) measurements on the fabricated sample were performed using a highly accurate and precise Keithley 4200 source measurement unit (SMU) inside a vacuum chamber equipped with a Janis probe station and Lakeshore temperature controller. IV measurements were performed in dark mode

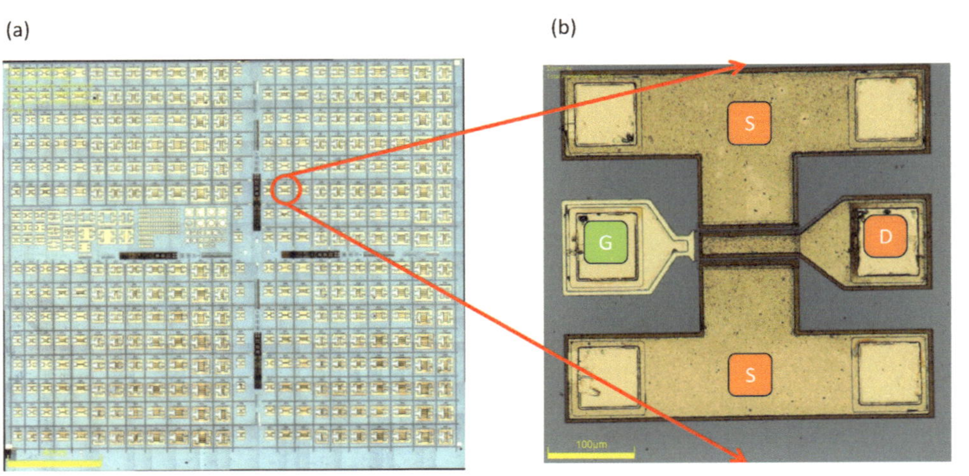

Figure 15.
(a) GaN HEMT fabricated wafer (b) 100°nm 2 × 100 GaN HEMT device (reprinted with permission from Ref. [56]).

(no illumination) and radiation illumination mode to extract the ISBT in fabricated GaN HEMT devices. 1-mW blue, yellow and red LEDs as well as 300-W halogen lamp-based perpendicular illumination sources were used in our experiment to excite the deep-level traps in the GaN HEMT device, while 1-mW broadband infrared illumination sources like a red laser (630–690 nm), near-infrared (NIR) LED (650–850 nm) and short-wave infrared (SWIR) LED (1.7–2.1 micron) based illumination at an oblique angle of incident (AOI) were used to investigate the ISBT at ambient temperature. Moreover, to confirm the transition is solely dependent upon the bandgap phenomenon, low-temperature PL and IV measurements were also carried out. Most devices showed the ambient temperature ISBT, however, we selected the 2 × 100 device for demonstration.

4. Results and discussion on experimental outcome

Post growth, the GaN heterostructure was extensively evaluated using standard, highly accurate semiconductor characterization techniques. **Table 3** shows the summary of extracted heterostructure properties using various characterization techniques. Room-temperature mobility and 2DEG carrier concentration were found to be 1885 $cm^2/V.s$ and 1.1E + 13 cm^{-2}, respectively, using the Hall measurement method. The composition and thickness of the AlGaN barrier layer play crucial roles in polarization and 2DEG carrier accumulation inside the GaN HEMT. Thickness of 21 nm and Al composition of 31% were found in the AlGaN barrier layer against targeted thickness of 20 nm and Al composition of 30%. Ohmic and Schottky contacts to GaN heterostructures play a vital role in the development of a GaN HEMT device. Low contact resistance of ~0.27 Ω.mm and high barrier height of ~0.72 eV were extracted using IV measurements. Surface traps were present in the GaN HEMT devices and led to significant degradation of DC and RF performance. High-quality Si_3N_4 surface passivation deposition was used to effectively reduce surface traps. The improvement in drain current density is about 35 mA/mm and in RF gain is 4 dB at 10 GHz after Si_3N_4 deposition, which clearly indicates the majority of surface traps are saturated after passivation. The saturation drain current density (@ Vg = 0 V) was measured at ~1 A/mm, while cut-off frequency of ~89 GHz was extracted for the fabricated 100-nm GaN HEMT.

FET shows the response towards THz beyond its cutoff frequency even at room temperature irrespective of semiconductor material systems [22, 24–28]. The Dyakonov–Shur plasma wave theory [22, 23] classically explains the THz behavior of the device starting from conventional semiconductors like Si, GaAs, and GaN to recently developed 2D materials system like graphene, MoS_2, WS_2, black phosphorous, and others. We proposed ISBT transition at ambient temperature as another

Properties	AlGaN	GaN	Measurement method
Thickness	21 nm	2.0	HR-XRD
Composition (%)	$Al_{0.31}Ga_{0.69}N$	—	HR-XRD
Lattice constant (A°)	5.121	5.185	HR-XRD
Bang gap (eV)	—	3.44	PL
2DEG	—	1.1E+13	Hall
Mobility (RT)	—	1885	Hall

Table 3.
Measured heterostructure properties using standard semiconductor characterization equipment.

potential mechanism for THz response of the GaN HEMT device. ISBT is demonstrated using IV measurement of the GaN HEMT device under dark and illuminated conditions. Usually, IV characteristics are the combination of all possible phenomenon in the FET. It is very difficult to distinguish the defects- or traps-assisted transitions, thermal transitions and ISBTs in the IV characteristics of FET. However, electrical tuning of ISBT in GaAs HEMT has been demonstrated [59]. There are three following key challenges involved in supporting the ambient ISBT mechanism in FET/HEMT.

1. rule out plasma wave mechanism

2. Defects–/traps-based transitions

3. Thermal energy-assisted transitions

4.1 Rule out plasma wave mechanism

The basic physics involved in plasma wave theory is that 2DEG instability in short-channel HEMTs has a resonant response to incident electromagnetic radiation. The resonance frequency is governed by the size and shape of the channel (i.e., the geometrical plasmon frequency). Tuning the plasmon resonant frequency to the incident THz wave is used for detectors, mixers and multipliers, as the carrier resonance happens in the THz frequency range only. It is not possible to generate plasma wave inside the FET channel if the incident radiation has a frequency other than THz. In other words, if we are using a source other than a THz radiation source that is capable of inducing the ISBT, the generation of plasma waves can be ruled out inside the FET/HEMT.

4.2 Defects/traps-based transitions

The deep-level traps- or defects-assisted transitions have been well reported since the invention of heterostructure [60]. The traps' energy level and density depend upon several parameters like heterostructure growth condition, materials system, and others. Especially in GaN-based wide bandgap semiconductor materials, the domination of the deep-level traps is even more significant than GaAs semiconductor material [61]. It is highly difficult to prevent the transitions through these traps. However, control over traps-based transition is possible, as it shows the different responses towards the incident radiations. If we are selecting the illumination source that has the least significance for trap excitation and the most significance for ISBT, then defects–/traps-assisted transitions can also be also ruled out.

4.3 Thermal energy-assisted transitions

The thermal energy associated at room temperature is $\sim$25 mev ($\sim$ 6 THz), which is much higher than the spacing between the subband in a quantum well. It is very difficult to negligible thermal energy contribution. Thermal occupation of electrons in a higher subband may prevent the observation of ISBT at ambient temperature [59]. Measurement are done at ambient as well as low temperature in vacuum condition with a precise and accurate temperature controller to quantified the thermal transitions. Furthermore, source-measurement units (SMUs) are accurate for detecting very small changes in measurement for dark and illuminated conditions. The background thermal energy contribution in transitions is equally present in both dark and illumination modes, which clearly indicates the presence of ISBT in the measurement.

In summary, to confirm the transitions solely occurring due to ISB inside the triangular quantum well of the heterostructure, we used a 1-mW SWIR LED because it is least significant for trap excitation [62, 63], whereas it is most significant for ISBT. Moreover, the use of a SWIR source that is not in the THz frequency range ensures that the generation of plasma wave inside the channel is not possible. The blue LED was selected for measuring traps-assisted transitions. **Table 4** summarizes the key challenges involved along with possible solutions to confirm room-temperature ISBT in GaN HEMT.

To excite the deep-level traps in a GaN heterostructure, 1-mW blue, yellow and red LEDs as well as a 300-W halogen lamp-based perpendicular illumination were used. It is well proven that as we move from NIR to UV radiations, the trap excitation becomes more efficient. It is difficult to excite traps larger than 870 nm [62, 63]. In our experiments, blue LED was found to be more efficient among all used light sources to excite the deep-level traps. To extract the trap-assisted transitions, a 90-degree AOI under blue LED illumination for 10 min was used. The Id-Vd characteristics and change in drain current (ΔId) of the 100-nm GaN HEMT without and with illumination are shown in **Figure 16(a)**. Deep-level traps-assisted transitions increased the drain current up to approximately 24 mA/mm as shown in **Figure 16(b)**. It was found that after 10 min of illumination, there was no further significant increase in drain current, which confirms that most traps were saturated and the equilibrium condition was reached.

Sr. No	Discrimination ISBT from other mechanism	Used excitation source/methods
1	Plasma wave mechanism	Non terahertz radiation source
2	Defects/traps induced transition	Blue LED
3	Thermal energy contribution	Measurement in vacuum, precise temperature control with highly accurate SMUs Confirm with low temperature IV and PL
4	ISBT	SWIR (1.7–2.1 μm) source

Table 4.
Measurement methods and excitation sources used to confirm ISBT.

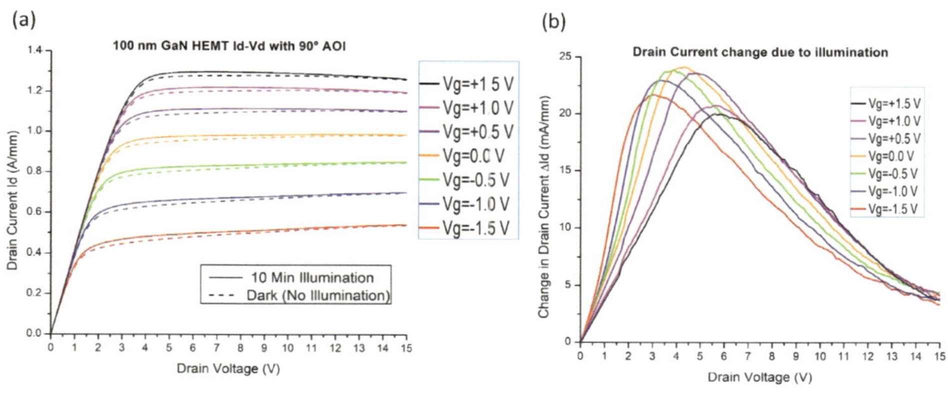

Figure 16.
Effect of 90° AO Iillumination with blue LED (a) on id-Vd characteristics of 100 nm GaN HEMT device (b) change in drain current (reprinted with permission from Ref. [56]).

Red laser (630–690 nm), NIR LED (650–850 nm) and SWIR LED (1.7–2.1 micron) broadband infrared sources were used in our experiment to investigate physical phenomena other than plasma wave at ambient temperature. It is noted that the ISB absorption characteristics were found to be identical for all used IR sources (red laser, NIR and SWIR LEDs) with the highest absorption found for the case of SWIR LED.

For ISBT experiments, we selected 1-mW SWIR LED as it is least significant for trap excitation [62, 63], whereas it is most significant for ISBT. GaN heterostructure materials have a wide bandgap with lower cut-off wavelengths than the wavelength of the IR light source, ensuring the transition of the carriers from valance band to the conduction band is forbidden.

When light is incident perpendicular to the sample surface ISBT cannot be induced, as the electric field has component only in the quantum-well plane [40]. We illuminated the sample at an oblique angle of incidence to discriminate ISBT with other transitions. When the sample is illuminated with an oblique angle, IR radiation interacts with carriers inside the subband of the triangular quantum well and transitions occur within the conduction band. The Id-Vd characteristics and change in drain current (ΔId) of the 100-nm GaN HEMT without and with 30 s of 45-degree AOI SWIR LED illumination are shown in **Figure 17(a)** and **(c)**. A zoom portion of the Id-Vd curve for −0.5 and − 1.0 gate voltage is shown in **Figure 17(b)** for visualization purposes, as the change in drain current was very small due to illumination. Infrared lamp-assisted photoinduced ISBT in doped and undoped multiple quantum wells was reported by Olszakier et al. in a series of experiments [64–68]. It was concluded that the ISBT involves free electrons as well as excitons. The exciton-based transitions have greater frequency and oscillator strength than those of the bare electrons.

The bulk wurtzite semiconductor band diagram along with the two E_0 and E_1 subbands in the triangular quantum well involves transition of free electrons and excitons-based transition as shown in **Figure 18(a)–(c)**, respectively. In the asymmetrical (triangular) quantum well, inversion symmetry with respect to the quantum well center is broken, which leads to a relaxation of the selection rules (i.e., transitions between all subbands are allowed) [40]. It is possible to tune subbands inside the quantum well by external electrical field in an HEMT device. Free electron-based ISBT (0.5–10 THz) and exciton-assisted ISBT (for higher frequency) can be exploited as potential tunable sources and detectors for the entire THz range.

The spacing between subband and quantum-well width depends on gate biasing. Let us consider only two subbands, E_0 and E_1 inside a well having N_0 and N_1 electrons, respectively. The gate voltage is selected in such a way where ground state E_0 is situated below the Fermi level as shown in **Figure 19(a)**. The 2DEG

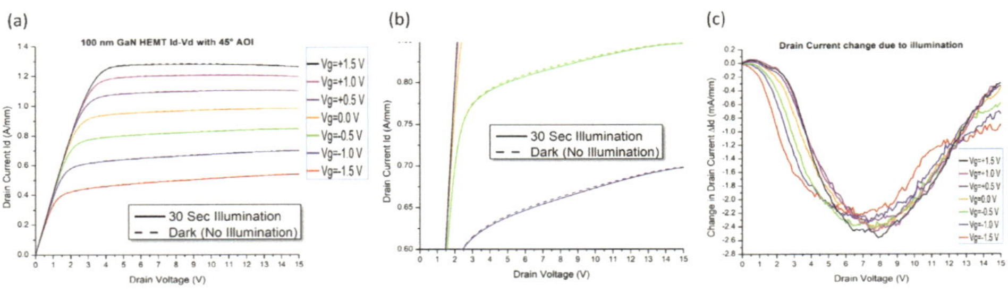

Figure 17.
Effect of 45° AO Iillumination with SWIR led (a) on id-Vd characteristics of 100°nm GaN HEMT device (b) zoom portion of id-Vd characteristics for drain current change visualization and (c) change in drain current (reprinted with permission from Ref. [56]).

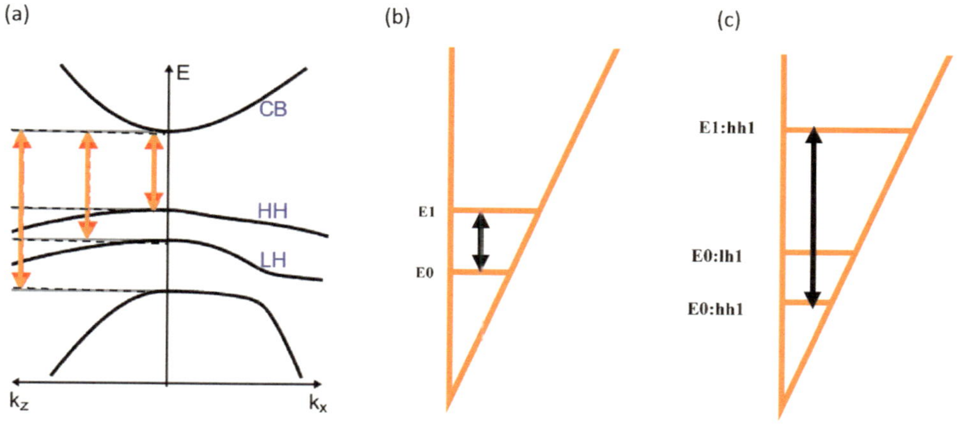

Figure 18.
(a) Band structure for wurtzite (WZ) bulk semiconductor with conduction band (CB), light and heavy holes (HH, LH). The ISBT is shown in (b) well electrons and (c) the exciton schemes. (reprinted with permission from Ref. [56]).

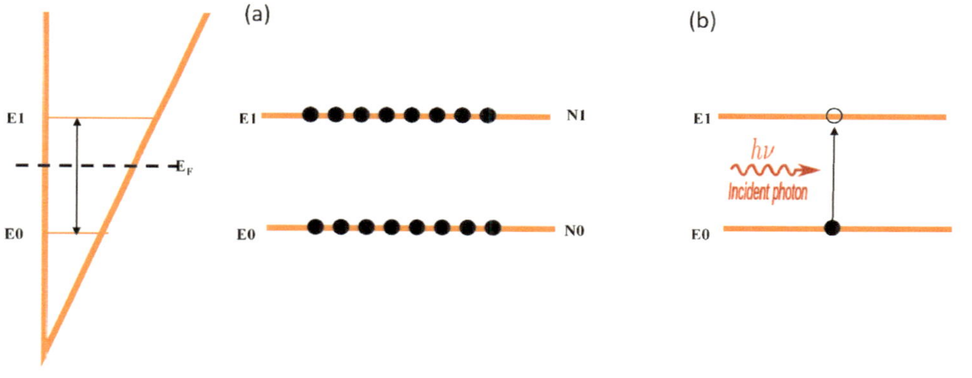

Figure 19.
(a) Band-schematic of the first two subbands in a 2DEG with respect to the Fermi level (b) absorption in the subband. (reprinted with permission from Ref. [56]).

carriers below the Fermi energy level are extracted as a drain current by applying a potential between source and drain. When the sample is illuminated, the electrons in a ground state E_0 interact with an external electromagnetic field. The electrons pick up photons from the illuminating field, which allows them to enter an excited energy state E_1 within the subband as shown in **Figure 19(b)**. These excited electrons are in the energy level E_1 that is above the Fermi level. As these electrons are not contributed to conduction, the drain current Id is decreased. This mechanism is clearly observed in **Figure 17(c)** in terms of decrease in drain current due to illumination, which shows ISB absorption. The amount of absorption strictly depends upon the distribution of electrons in the subband and the spacing between subband and width of well. To rule out thermal energy contribution in IV characteristics, measurement is done in vacuum conditions. The precise and accurate temperature controller and SMUs are used in measurement, which are able to detect a very small change in drain current in dark and illuminated conditions. Moreover, to confirm the transition is solely dependent upon the bandgap phenomenon, low-temperature PL and IV measurements were carried out. The temperature-dependent bandgap shifting in GaN found in PL measurement, as shown in

Figure 20, matches with previously published results [69]. Low-temperature 200-K and 100-K ISB absorption measurements were also carried out. It was found that the intensity of absorption increases as the temperature decreases, as shown in **Figure 21(a)** and **(b)**. It indicates that thermal energy contribution decreases with a decrease in temperature. The temperature-dependent bandgap variation in GaN perfectly matches with ISB absorption (Vg = 0 V, Vds = 8 V), as depicted in **Figure 22**.

In conclusion, low-temperature and angle-dependent illumination-based measurements were used to confirm the ISB transition in GaN HEMT. We have

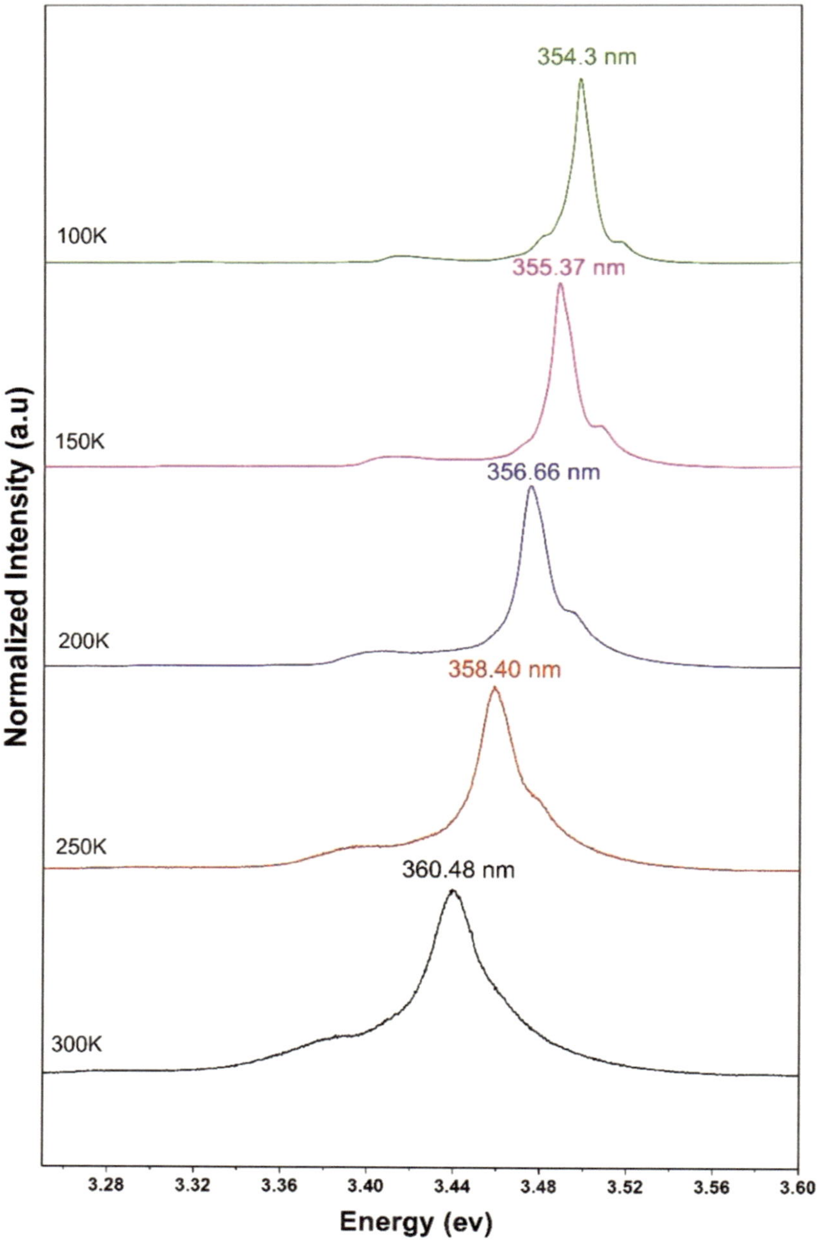

Figure 20.
Low-temperature PL measurement of GaN heterostructure (reprinted with permission from Ref. [56]).

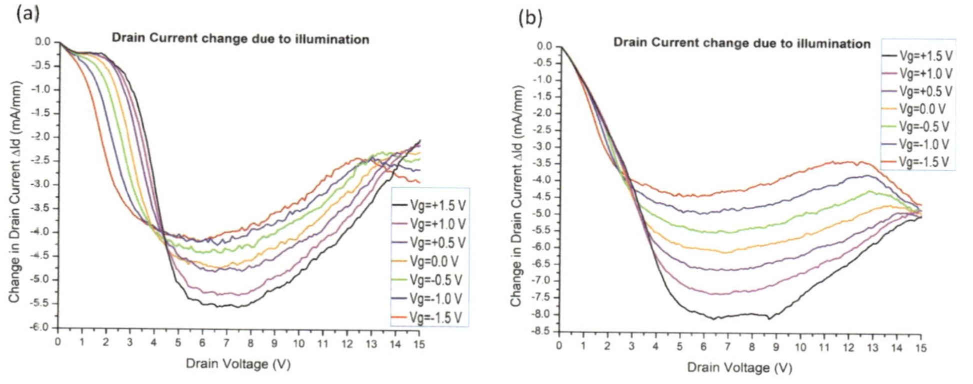

Figure 21.
Change in drain current due to 45° AOI and 30°second illumination with SWIRLED at temperature (a) 200 K and (b) 100 K (reprinted with permission from Ref. [56]).

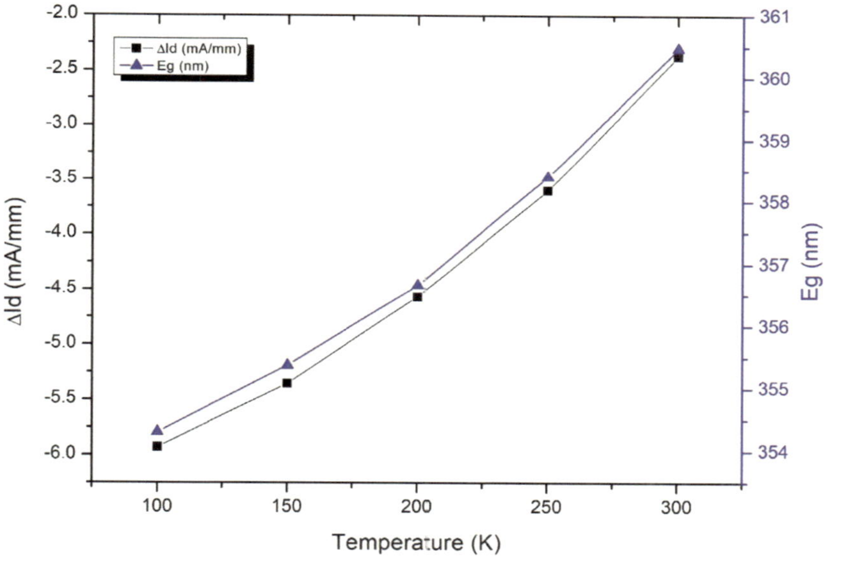

Figure 22.
Temperature dependent GaN band gap and change in drain current (vg = 0 V and Vds = 8 V) due to illumination (reprinted with permission from Ref. [56]).

experimentally explored electrical tuning of ISB resonance phenomena inside the triangular quantum well for a GaN HEMT device, which shows the potential of GaN HEMT technology to be realized as a room-temperature THz source and detector.

5. Conclusion

We have developed theoretical models for electrically tunable plasmonic metamaterials-assisted ISBT in GaN HEMT. Experimental demonstration of electrical tuning of ISBT in a GaN HEMT device at room temperature has not only provided a new alternate mechanism but also discriminates ISBT from other transitions induced by deep-level traps and defects in the 100-nm GaN HEMT device. The chapter also explored the photonics ISBT phenomenon in a GaN HEMT device for external biasing, which depends on tuning of the subband. A novel approach for ISBT in GaN HEMT helps to overcome the THz gap in the electromagnetic spectrum at ambient temperature.

Acknowledgements

We are thankful to the director of SAC for continuous encouragement and guidance during this study. We extend our sincere thanks to Prof Solomon Ivan at the Department of Physics, IIST Thiruvananthapuram, for helpful discussions and providing valuable suggestions. We are also thankful to the Microelectronics Group for providing fabrication and characterization support.

Author details

Rakesh Kaneriya[1,2]*, Gunjan Rastogi[1], Palash Basu[3], Rajesh Upadhyay[1] and Apurba Bhattacharya[1]

1 Microelectronics Group, Space Applications Centre, ISRO, Ahmedabad, India

2 Department of Physics, Indian Institute of Space Science and Technology, Thiruvananthapuram, India

3 Department of Avionics, Indian Institute of Space Science and Technology, Thiruvananthapuram, India

*Address all correspondence to: rakeshk@sac.isro.gov.in

References

[1] Tonouchi M. Cutting-edge THz technology. Nature Photonics. 2007;**1**: 97-105. DOI: 10.1038/nphoton.2007.3

[2] Xiang Yang, Xiang Zhao, Ke Yang, Yueping Liu, Yu Liu, Weiling Fu, Yang Luo. Biomedical applications of terahertz spectroscopy and imaging. Trends in Biotechnology. **34**(10);2016: 810-824

[3] Mittleman DM. Twenty years of terahertz imaging. Optics Express. 2018; **26**:9417-9431

[4] Rahman A, Rahman AK, Rao B. Early detection of skin cancer via terahertz spectral profiling and 3D imaging. Biosensors & Bioelectronics. 2016;**82**: 64-70

[5] Nagai N, Sumitomo M, Imaizum M, Fukasawa R. Characterization of electron- or proton irradiated Si space solar cells by THz spectroscopy. Semiconductor Science and Technology. 2006;**21**:201-209

[6] Nagai N, Imai T, Fukasawa R, Kato K, Yamauchi K. Analysis of the intermolecular interaction of nanocomposites by THz spectroscopy. Applied Physics Letters. 2004;**85**: 4010-4012

[7] Siegel P. THz instruments for space. IEEE Transactions on Antennas and Propagation. 2007;**55**(11):2957-2965

[8] Siegel PH. THz for space: The golden age. 2010 IEEE MTT-S International Microwave Symposium, 2010. pp. 816-819. DOI: 10.1109/MWSYM.2010.5515761

[9] Ferguson B, Zhang XC. Materials for terahertz science and technology. Nature Materials. 2002;**1**:26-33. DOI: 10.1038/nmat708

[10] Khanal S, Zhao L, Reno JL, Kumar S. Temperature performance of terahertz quantum-cascade lasers with resonant-phonon active-regions. Journal of Optics. 2014;**16**:094001. DOI: 10.1088/2040-8978/16/9/094001

[11] Rastogi G, Kaneriya RK, Sinha S, Upadhyay RB, Bhattacharya AN. Physics based simulation for studying the impact of contact resistance on DC and RF characteristics of AlGaN/AlN/GaN HEMT. Journal of Radio Science. 2019;**54**:904

[12] Kaneriya RK, Rastogi G, Basu PK, Upadhyay RB, Bhattacharya AN. Intersubband device modeling of gallium nitride high electron mobility transistor for terahertz applications. Journal of Radio Science. 2019;**54**:1172-1180. DOI: 10.1029/2019RS006844

[13] Kaneriya RK, Rastogi G, Basu PK, Upadhyay RB, Bhattacharya AN. Modeling of Electrically Tunable Metamaterial Embedded Intersubband Transitions in GaN HEMT for Terahertz Applications. 2020 URSI Regional Conference on Radio Science (URSI-RCRS), 2020, pp. 1-5, doi: 10.23919/URSIRCRS49211.2020.9113510

[14] Smorchkova IP, Chen L, Mates T, Shen L, Heikman S, et al. AlN/GaN and (Al,Ga)N/AlN/GaN two-dimensional electron gas structures grown by plasma-assisted molecular-beam epitaxy. AIP Publishing LLC. Journal of Applied Physics. 2001;**90**:5196. DOI: 10.1063/1.1412273

[15] Shen L, Heikman S, Moran B, Coffie R, Zhang N-Q, Buttari D, et al. AlGaN/AlN/GaN high-power microwave HEMT. IEEE Electron Device Letters. 2001;**22**:10

[16] Bernardini F, Fiorentini V, Vanderbilt D. Spontaneous polarization and piezoelectric constants of III-V nitride. Physical Review B. 1997;**56**:10024

[17] Weiwei K. TCAD simulation and modeling of AlGaN/GaN HFET [thesis]. 2008, North Carolina State University, Raleigh, North Carolina

[18] Silvaco® Application Note. State of the art 2D and 3D process and device simulation of GaN-based devices. A Journal for Process and Device Engineers. 2013;**23**(3):1-11

[19] Silvaco® Application Note. Atlas Simulation of GaN-Based Super Heterojunction Field Effect Transistors Using the Polarization Junction Concept, Simulation Standard. 2014

[20] Albrecht JD, Wang RP, Ruden PP, Farahmand M, Brennan KF. Electron transport characteristics of GaN for high temperature device modeling. Journal of Applied Physics. 1998;**83**(9):4777-4781. DOI: 10.1063/1.367269

[21] Yue Y, Hu Z, Guo J, Rodriguez BS, Li G, Wang R, et al. InAlN/AlN/GaN HEMTs with regrown ohmic contacts and fT of 370 GHz. IEEE Electron Device Letters. 2012;**33**(7):988-990. DOI: 10.1109/LED.2012.2196751

[22] Dyakonov M, Shur M. Shallow water analogy for a ballistic field effect transistor: New mechanism of plasma wave generation by dc current. Physical Review Letters. 1993;**71**(15): 2465-2468. DOI: 10.1103/PhysRevLett.71.2465

[23] Dyakonov M, Shur M. Detection, mixing, and fequency multiplication of terahertz radiation by two-dimensional electronic fluid. IEEE Transactions on Electron Devices. 1996;**43**(3):380-387. DOI: 10.1109/16.485650

[24] Nahar S, Shafee M, Blin S, Penarier A, Nouvel P, Coquillat D, et al. Wide modulation bandwidth terahertz detection in 130 nm CMOS technology Eur. Journal of Applied Physics. 2016; **76**:20101. DOI: 10.1051/epjap/2016160302

[25] Hou HW, Liu Z, Teng JH, Palacios T, Chua SJ. High temperature terahertz detectors realized by a Gan high electron mobility transistor. Scientific Reports. 2017;7:46664. DOI: 10.1038/srep46664

[26] Kurita Y, Ducournau G, Coquillat D, Satou A, Kobayashi K, Tombet SB, et al. Ultrahigh sensitive sub-terahertz detection by InP-based asymmetric dual-grating-gate high-electron-mobility transistors and their broadband characteristics. Applied Physics Letters. 2014;**104**:251114. DOI: 10.1063/1.4885499

[27] Wang L, Chen X, Yu A, Zhang Y, Ding J, Lu W. Highly sensitive and wide-band tunable terahertz response of plasma waves based on graphene field effect transistors. Scientific Reports. 2014;**4**:5470. DOI: 10.1038/srep05470

[28] Viti L, Hu J, Coquillat D, Politano A, Knap W, Vitiello MS. Efficient terahertz detection in black-phosphorus nano-transistors with selective and controllable plasma-wave, bolometric and thermoelectric response. Scientific Reports. 2016;**6**:20474. DOI: 10.1038/srep20474

[29] Sun JD, Sun YF, Wu DM, Cai Y, Qin H, Zhang BS. High-responsivity, low-noise, room-temperature, self-mixing terahertz detector realized using floating antennas on a GaN-based field-effect transistor. Applied Physics Letters. 2012;**100**:013506. DOI: 10.1063/1.3673617

[30] Javadi E, Delgado-Notario JA, NMasoumi MS, Elazquez-Perez VJE, Meziani YM. Continuous wave terahertz sensing using GaN HEMTs. Physica Status Solidi (a) 215 (2018): 1700607

[31] Hou HW, Liu Z, Teng JH, Palacios T, Chua SJ. A sub-terahertz broadband detector based on a GaN high-electron-mobility transistor with nanoantennas.

Applied Physics Express. 2017;**10**: 014101. DOI: 10.7567/APEX.10.014101

[32] Fatimy A, Dyakonova N, Meziani Y, Otsuji T, Knap W, Vandenbrouk S, et al. AlGaN/GaN high electron mobility transistors as a voltage-tunable room temperature terahertz sources. Journal of Applied Physics. 2010;**107**:024504. DOI: 10.1063/1.3291101

[33] Chuang SL. Physics of Optoelectronic Devices. New York: John Wiley and Sons; 1995

[34] Chuang SL. Physics of Photonics Devices. New York: John Wiley and Sons; 2009

[35] Bradley AF. Envelope-function formalism for electrons in abrupt heterostructures with material-dependent basis functions. Physical Review B. 1996;**54**:1909-1921

[36] Eissfeller T, Vogl P. Real-space multiband envelope-function approach without spurious solutions. Physical Review B. 2011;**84**(19):195122. DOI: 10.1103/PhysRevB.84.195122

[37] Foreman BA. Effective mass Hamiltonian and boundary conditions for the valence bands of semiconductor microstructures. Physical Review B. 1993;**48**(7):4964-4967. DOI: 10.1103/PhysRevB.48.4964

[38] Wieck AD, Thiele F, Merkt U, Ploog K, Weimann G, Schlapp W. Subband-landau-level coupling in GaAs/Ga1−xAlx As heterojunctions. Physical Review B. 1989;**39**(6):3785-3794. DOI: 10.1103/PhysRevB.39.3785

[39] Ando T, Fowler AB, Stern F. Electronics properties of two dimensional systems. Reviews of Modern Physics. 1982;**54**(2):437–672. DOI: 10.1103/RevModPhys.54.437

[40] Helm M. The basic physics of intersubband transitions. In: Liu H, Capasso F, editors. Intersubband Transitions in Quantum Wells: Physics and Device Applications I, Semiconductors and Semimetals. Vol. 62. Academic Press; San Diego, CA, USA; 1999. pp. 20, 26

[41] Scalari G, Maissen C, Hagenmüller D, De Liberato S, Ciuti C, Reichl C, et al. Ultrastrong light-matter coupling at terahertz frequencies with split ring resonators and inter-Landau level transitions. Journal of Applied Physics. 2013;**113**:136510

[42] Benz A, Montaño I, Klem JF, Brener I. Tunable metamaterials based on voltage controlled strong coupling. Applied Physics Letters. 2013;**103**: 263116

[43] Scalari G, Maissen C, Turcinková D, Hagenmüller D, De Liberato S, Ciuti C, et al. Ultrastrong coupling of the cyclotron transition of a 2D electron gas to a THz metamaterial. Science. 2012; **335**:1323

[44] Gabbay A, Brener I. Theory and modeling of electrically tunable metamaterial devices using intersubband transitions in semiconductor quantum wells. 2012 Optical Express 20(6):6548

[45] Benz A, Campione S, Liu S, Montano I, Klem JF, Sinclair MB, et al. Monolithic metallic nanocavities for strong lightmatter interaction to quantum-well intersubband excitations. Optical Express. 2013; **21**(26):32572

[46] Shrekenhamer D, Rout S, Strikwerda A, Bingham C, Averitt RD, Sonkusale S, et al. High speed terahertz modulation from metamaterials with embedded high electron mobility transistors. Optical Express. 2011; **19**(10):9968

[47] Zhang Y, Zhao Y, Liang S, Zhang B, Wang L, Zhou T, et al. Large phase

modulation of THz wave via an enhanced resonant active HEMT metasurface. Nano. 2018;**8**(1):153

[48] Vicarelli L, Vitiello MS, Coquillat D, Lombardo A, Ferrari AC, Knap W, et al. Graphene field-effect transistors as room-temperature THz detectors. Nature Materials. 2012;**11**:865

[49] Zak A, Andersson M, Bauer M, Matukas J, Lisauskas A, Roskos H, et al. Antenna-integrated 0.6 THz FET direct detectors based on CVD graphene. Nano Letters. 2014;**14**(10):5834

[50] Tong J, Muthee M, Chen S, Yngvesson S, Yan J. Antenna enhanced graphene THz emitter and detector. Nano Letters. 2015;**15**(8):5295

[51] Viti L, Hu J, Coquillat D, Politano A, Knap W, Vitiello MS. Efficient terahertz detection in black-phosphorus nano-transistors with selective and controllable plasma-wave, bolometric and thermoelectric response. Scientific Reports. 2016;**6**:20474

[52] Berry C, Wang N, Hashemi M, Unlu M, Jarrahi M. Significant performance enhancement in photoconductive terahertz optoelectronics by incorporating plasmonic contact electrodes. Nature Communications. 2016;**4**:1622

[53] Tanoto H, Teng J, Wu Q, Sun M, Chen Z, Maier S, et al. Greatly enhanced continuous-wave terahertz emission by nano-electrodes in a photoconductive photomixer. Nature Photonics. 2012;**6**:121

[54] Wang N, Cakmakyapan S, Lin Y, Javadi H, Jarrahi M. Room-temperature heterodyne terahertz detection with quantum-level sensitivity. Nature Astronomy. 2019;**3**:977

[55] Yu N, Wang Q, Kats M, Fan J, Khanna S, Li L, et al. Designer spoof surface plasmon structures collimate terahertz laser beams. Nature Materials. 2010;**9**:730

[56] Kaneriya RK, Rastogi G, Basu PK, Upadhyay RB, Bhattacharya AN. Room temperature photon induced electrical tuning of intersubband transition in GaN HEMT for terahertz applications. Microelectronic Engineering. 2020;**233**: 111433

[57] Rastogi G, Kaneriya R, Sinha S, Upadhyay R. Optimization of ohmic contact fabrication for Al0.3Ga0.7N/AlN/GaN HEMTs on 6H-SiC using recess etching and surface plasma treatment processes. Journal of Nanomaterials & Molecular Nanotechnology. 2019;**8**:2. DOI: 10.4172/2324-8777.1000267

[58] Rastogi G, Kaneriya R, Upadhyay R, Bhattacharya AN. Optimization of Ni/Au Schottky contacts on Al0.3Ga0.7N/AlN/GaN heterostructure for RF application. International Journal of Nanotechnology and Nanoscience. 2020;**6**:8-15

[59] Pal S, Valentin S, Ludwig A, Wieck A. Quantum Confinement in High Electron Mobility Transistors. IntechOpen, London, UK; 2017. DOI: 10.5772/intechopen.68374

[60] Mooney PM. Deep donor levels (DX centers) in III-V semiconductors. Journal of Applied Physics. 1990;**67**(3): R1. DOI: 10.1063/1.345628

[61] Polyakov AY, Lee IH. Deep traps in GaN-based structures as affecting the performance of GaN devices. Materials Science and Engineering. 2015;**94**:1-56. DOI: 10.1016/j.mser.2015.05.001

[62] Liang Y, Jia L, He Z, Fan Z, Zhang Y, Yang F. The study of the contribution of the surface and bulk traps to the dynamic Rdson in AlGaN/GaN HEMT by light illumination.

Applied Physics Letters. 2016;**109**: 182103. DOI: 10.1063/1.4966536

[63] Polyakov A, Smirnov N, Shchemerov I, Lee H, Jang T, Dorofeev A, et al. Current relaxation analysis in AlGaN/GaN high electron mobility transistors. Journal of. Vacuum Science and Technology B. 2017;**35**:011207-1 to 10. DOI: 10.1116/1.4973973

[64] Olszakier M, Ehrenfreund E, Cohen E, Bajaj J, Sullivan GJ, Miller D. Intersubband spectroscopy in GaAs/GaAlAs multi-quantum-well structures: Photoinduced absorption. Superlattices and Microstructures. 1989;**5**:283. DOI: 10.1016/0749-6036(89)90301-7

[65] Olszakier M, Ehrenfreund E, Cohen E, Bajaj J, Sullivan GJ. Photoinduced intersubband absorption in undoped multi-quantum-well structures. Physical Review Letters. 1989;**62**:2997. DOI: 10.1103/PhysRevLett.62.2997

[66] Olszakier M, Ehrenfreund E, Cohen E, Bajaj J, Sullivan GJ. Intersubband absorption by photoinduced excitons in undoped multi quantum wells. Surface Science. 1989;**228**:123-126. DOI: 10.1016/0039-6028(90)90272-A

[67] Garini Y, Olszakier M, Cohen E, Ehrenfreund E, Ron A, Law KK, et al. Photoinduced intersubband absorption in barrier doped multi-quantum-wells. Superlattices and Microstructures. 1990; **7**:287-290. DOI: 10.1016/0749-6036(90)90211-O

[68] Cohen E, Ehrenfreund E, Garini Y, Olszakier M, Ron A. Photo-induced intersubband transitions in quantum wells. In: Rosencher E et al., editors. Lntersubband Transitions in Quantum Wells. New York: Plenum Press; 1992. p. 264

[69] Romero MF, Feneberg M, Moser P, Berger C, Bläsing J, Dadgar A, et al. Luminescence from two-dimensional electron gases in InAlN/GaN heterostructures with different In content. Applied Physics Letters. 2012; **100**:212101. DOI: 10.1063/1.4720087

Chapter 9

Interdigitated Photoconductive Antenna for Efficient Terahertz Generation and Detection

Shyamal Mondal, Nisha Flora Boby Edwin and Vaisshale Rathinasamy

Abstract

THz signals can be generated commonly from Photoconductive Antenna (PCA) but the efficiency is low for the conventional PCA. This work improves the optical to terahertz conversion efficiency of the terahertz radiation by changing the conventional PCA structure to Interdigitated PCA (IPCA). The efficiency of PCA is dependent on the current pulse generated in the antenna structure when the laser pulse is incident on it. This paper targets to achieve high photo-current, as well as THz electric field from the IPCAs which are simulated using FEM and FDTD techniques. Also, the effect of various parameters such as current, gain, frequency bandwidth, optical to terahertz conversion efficiency, etc. are studied to study the importance of IPCAs.

Keywords: interdigitated PCA, light-matter interaction, millimeter wave, photoconductive antenna, terahertz

1. Introduction

The current research and development of terahertz (THz) and millimeter wave technology at a global scale demonstrate numerous applications in medical imaging, security, high-speed communication, material characterization and spectroscopy [1]. Such applications increase the demands for efficient THz sources and detectors. Broad spectra are useful for many applications, such as time-domain spectroscopy, multi-input multi-output (MIMO) communication, etc. Photoconductive antennas (PCA) have been widely accepted as a reliable source and detector for THz generation and detection. It provides advantages for optically producing and detecting THz radiation [2]. However, there is the issue of low optical-to-THz conversion efficiency [3, 4]. Efforts have been made to enhance the efficiency by improving the laser pulse coupling, including the use of anti-reflection coating on Low- Temperature-grown Gallium Arsenide (LT-GaAs) [5], AlAs-AlGaAs based Bragg reflector under the LT-GaAs layer [6], nanoplasmonic structures [7], nanoplasmonic double layer structure [8, 9], recessed electrode and recessed nanoplasmonic array, nano-spaced electrodes [10], optical plasmonic nano-antenna [11], plasmonic nanostructure [12], graphene [13].

The design of an efficient photoconductive antenna requires a thin film of a highly resistive direct semiconductor material (III–V group) placed over the

IntechOpen

substrate and a pair of electrodes. Low Temperature grown-Gallium Arsenide based thin film and Semi Insulating—Gallium Arsenide based substrate are normally used. The THz output power (or electric field), spectral bandwidth and optical-to-THz conversion efficiency of a PCA highly depend on its geometry, dimensions and input laser parameters [14]. Based on the aperture gap between the anode and cathode, PCAs can be classified into small gap, semi-large gap and large gap types. To improve the performance, different PCA geometries and arrays have been reported [15]. In the literature, a bow-tie PCA structure provides frequency-independent characteristics, bandwidth and power, which are widely used for THz generation [16, 17]. Experimental analysis of THz far-field radiation for a butterfly-shaped PCA was conducted by the researchers in [18]. Recent publications [19] enhance THz radiation with the ZnO nanorods, which acts as a concentrator and an anti-reflector. The increased photocurrent comes the increased local fields and the decreased backward reflection of the optical pump.

This chapter explores the prospects of an inter-digitated PCA (IPCA) for THz generation and detection. Compared to the conventional large gap dipole PCA or bow-tie PCA, the IPCA geometry shows better performance in generating THz pulses because it takes the advantages from both the small gap and large gap PCA. As the gap between the PCA electrodes is filled by the metal teeth-like structures, most of the generated photocarriers get collected at the respective electrodes with less carrier drift time and leads to the uniform electric field. The addition of slots, the number of teeth-like electrodes and teeth width dimension have been varied to study the further improvement on IPCA based THz pulse generation [20]. The IPCA with slots helps to shift the resonant frequency toward higher THz frequencies. The slots have been placed at different positions and optimized to study geometry-dependent THz signals. The slotted IPCA and IPCA modeling, simulation, experimental results and its applications have been discussed in the following sections.

2. THz generation and detection using PCA

Some of the electronic sources to generate terahertz radiation are vacuum and solid-state devices, such as the gyrotron, backward wave-oscillators, traveling wave tubes [21]. These devices are bulky and require high magnetic field. The nonlinear optical properties are exhibited for THz wave generation by nonlinear crystals like ZnTe, GaP, InGaAs via difference frequency generation (DFG) and optical rectification (OR) methods [1]. In these optical radiated THz sources, the thickness of the nonlinear material should be carefully chosen and high THz power can be generated by the high optical input power. The limitations are phase matching between the optical field and THz. Based on the optical excitation type, the THz antennas are classified into THz PCA in the pulsed system and THz photomixers in continuous wave systems. The continuous wave systems use two different frequency monochromatic lasers. The THz radiated power can be improved by enhancing the photocurrent inside the semiconducting material and also by modifying the structure of PCA. Another approach to generate the continuous THz signals is photo mixing or optical heterodyne conversion which is achieved by mixing two monochromatic continuous-wave lasers [22]. These two lasers with frequencies ω_1 and ω_2 and phases of β_1 and β_2 respectively produce beating and modulate the photoconductive switch conductance at the THz difference frequency. The combined beam is targeted on the electrodes and generate terahertz radiation with ω_0 frequency. Similar to the THz PCA, the photomixers also consists of photoconductive material, metal electrode [23]. The laser diode is used as the input optical source for the photomixers and this has advantages like compact in structure, less complexity and lightweight.

To generate THz pulsed signals, the photoconductive antenna can be used both as an emitter and detector based on ultrafast optical techniques. The photoconductive antenna shows better performance in all aspects of terahertz generation based on the photoconduction principle [24]. In terahertz pulsed systems, the DC bias voltage is applied and the photocarriers in the semiconductor photoconductive material give rise to current density and ultrafast THz pulses [25]. The photoconductive antenna is referred to as Auston switches similar to the Hertzian dipole structure discovered in 1984 by Auston [26]. This PCA is structurally similar to the RF/MW antennas. The differences are the semiconductive materials. Silicon, InGaAs, GaAs and low-Temperature grown-GaAs (LT-GaAs) are used as the substrates instead of the dielectric substrate materials [27, 28]. The PCA consists of the metal electrodes on the photoconductive material substrate with an optical laser source and bias voltage. The metal electrodes made of silver, gold, aluminum, etc., can also be used as biasing pads. This works with the photoconduction principle. The terahertz PCA can be classified into small gap antenna, semi-large gap antenna and large gap antenna based on the antenna gap size.

The performance of PCA gets affected by the geometry of the antenna, optical source, impedance matching. Some antenna structures like bow-tie PCA, dipole PCA, logarithmic spiral antenna, nanoplasmonic PCA [29], nanoantennas with plasmonic contact electrode gratings, Amplifier-driven large-area PCA, Schottky PCA, Four contact PCA, split ring resonators [30] etc., can be used as the terahertz sources and detectors. To enhance the performance of photoconductive antenna in THz radiation, the antenna structures can be optimized. The sharp edges of the electrodes are very important to produce high electric-field but its fabrication is difficult and also there is some restriction phenomenon reducing the THz photocurrent generation which are velocity overshoot phenomenon, screening effects, etc., [25, 31]. Among these designs, the interdigitated PCA shows better performance in terahertz generation and detection. Hence, the slotted IPCA and IPCA designs has been explored in next section.

3. Interdigitated PCAs

In the terahertz research area, the THz waves with high SNR and large spectral bandwidth can be emitted by using innovative antenna geometry designs. Even though the conventional PCAs have many advantages over THz generation, there are the following drawbacks as well, (i) a few photocarriers can reach the corresponding electrodes due to large antenna gap area which leads to the screening effect [32], (ii) more photo-carriers can destabilize the THz output due to thermal effect, (iii) THz output power saturates quickly with the increase of laser power, and (iv) less THz output power, hence, low optical-to-THz conversion efficiency. To overcome these disadvantages, the interdigitated photoconductive antenna (IPCA) geometry can be used. The interdigitated photoconductive antenna emits the pulsed THz wave and the interdigitated photomixers can be used to generate continuous THz waves. This structure combines both the advantages of a large and small gap antenna. The finger-like electrodes are included in the antenna gap area which helps in reducing the active area and reduces the carrier drift time. This antenna structure requires less input power to perform high E-fields and high SNR compared to other PCA designs. The THz output beam always exits the surface of the substrate in a cone and diverges very fast from the propagation direction. Usually, metal parabolic mirrors and high-index THz lenses are used with the IPCA. They help to collimate and re-focus the THz beam. The other techniques such as second metallization [33], micro-lens array [34], binary phase masking [35] are also used in integrated THz devices to have highly directional output beam.

By varying the number of metal electrodes in the IPCA gap and keeping the active area and antenna gap area constant in **Figure 1**a, the optical to terahertz conversion efficiency has been improved [21] compared to the conventional PCA. The optical to terahertz conversion efficiency of conventional PCA (dipole PCA) is very low (0.00075%) [36]. With interdigitated structures reported in [20, 37–39], the optical to terahertz conversion efficiency has been improved from 0.000614 to 0.0678% by increasing the number of interdigitated teeth from 2 elements to 40 elements. These interdigitated PCA structures can also be included in the bow-tie and dipole PCA gap area. **Figure 1** shows the schematic diagram of the interdigitated PCA, IPCA in bow-tie PCA and tip-to-tip PCA. All these PCAs have a LTg-GaAs layer of a 300 μm length, a 300 μm width and a 30 μm thickness. The metallic structures are made of silver with a 1 μm thickness. The Length of the teeth-like electrodes are also kept constant as L_e = 21 μm. The gap between the teeth-like electrodes are kept constant for all the design as S_g = 1 μm. The interdigitated PCA shown in **Figure 1a** has been designed for teeth width of 21.5 μm (i.e. 2-elements), 8 μm (i.e. 5-elements), 2.46 μm (i.e. 13-elements) and 100 nm (i.e. 40-elements). **Figure 1b** shows the IPCA is included in the conventional bow-tie PCA. Another type of IPCA is the tip-to-tip PCA (**Figure 1c**). The antenna design consists of cathode and anode electrode fingers like comb structures [40, 41]. The electric field distribution becomes strong in the middle part of the photocond uctive antenna. The tip-to-tip PCA fingers are rectangular or trapezoidal shape. The trapezoidal fingers produce a stronger electric field magnitude than that of the rectangular ones [25]. Using these tip-to-tip electrodes, the large capacitance can be mitigated by

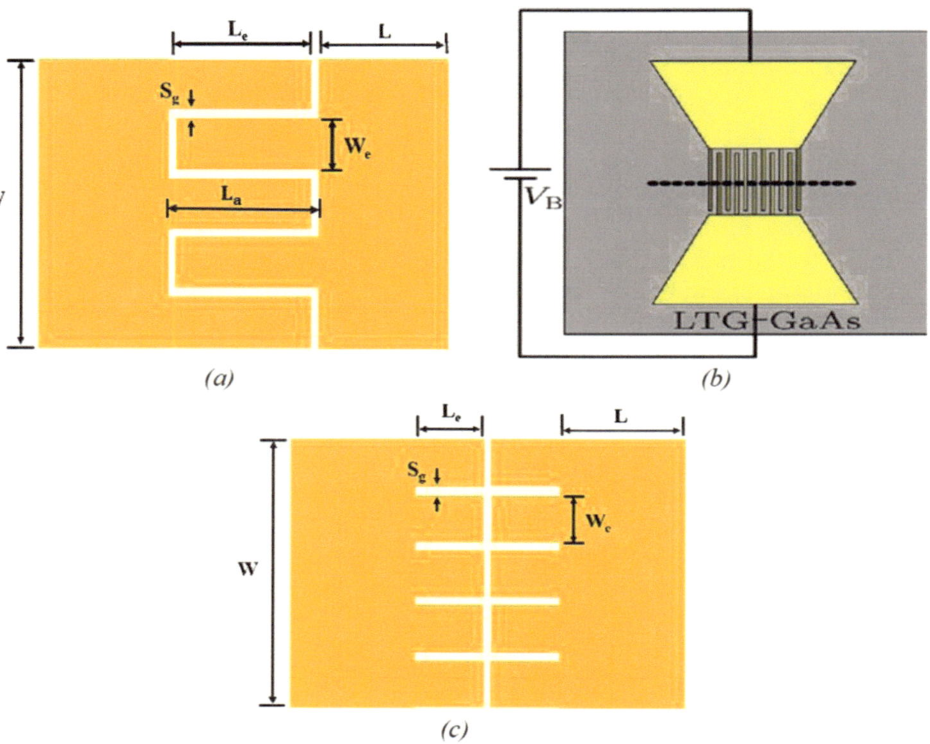

Figure 1.
Schematic representation of (a) interdigitated PCA (b) IPCA in bow-tie PCA and (c) tip-to-tip PCA. The parameters L is the length of the electrode-pad, L_e is the electrode teeth length, W is the width of the electrode-pad, W_e is the width of the electrode teeth, S_g is the gap between the electrodes, S_l is the slot length and S_w is the slot width.

extending the active area of fingers. Compared with the conventional PCA, this IPCA design produces enhanced THz E-field and short carrier drift time between the electrodes. In pulsed system, the high THz radiated power can be obtained by using this efficient geometry [40–42].

4. Slotted IPCA

The slotted IPCA is another kind of interdigitated PCA. The ultrashort femtosecond laser pulse is illuminated at the centre part of the antenna. The IPCA consists of a highly defect photoconductive material LT-GaAs as substrate of a length $L_s = 300$ μm, a width $W_s = 300$ μm and a 3 μm as thickness t. The planar silver electrode of a 1 μm thickness is placed on the substrate. The periodic metallic structures are placed at the edges of the anode and cathode with a gap of 1 μm, teeth width of W_e = 8 μm and a teeth length of L_e = 20 μm. These slots in the electrodes improve the field intensity and spectrum bandwidth for strong THz radiation [37]. By varying the geometry of interdigitated finger and slots, the performance of IPCA gets improved. **Figure 2a** shows the IPCA with two parallel slots S_l = 40 μm and S_w = 2 μm. **Figure 2b** and **c** show the IPCA with a center slot and two side slots, where L_w = 20 μm, S_l = 20 μm and S_w = 2 μm. The length and width of the teeth-like electrodes and substrate are consistent to those in **Figure 1**. The additional slots in the IPCA improves the performance of THz radiation. The center-slot realizes the maximum peak intensity of about 38.59 W cm^{-2}. If two parallel slots are added, the peak intensity is reduced to 27.20 W cm^{-2}. If the slots are added around the center, the peak intensity reduces to 26.44 W cm^{-2} [20]. This geometry generates high THz radiated power by optimizing the length and width of

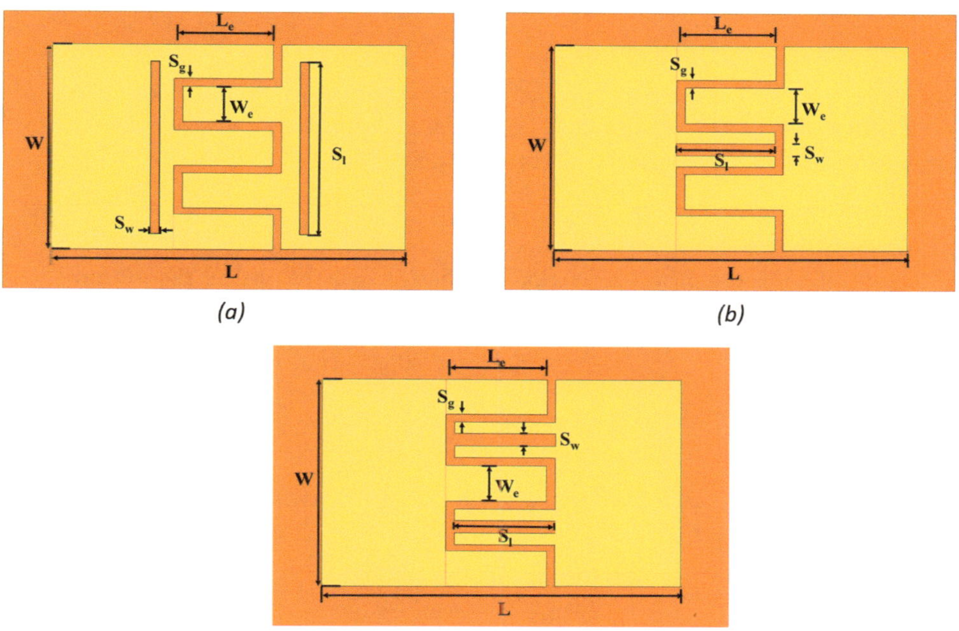

Figure 2.
Top view of slotted IPCA with (a) two parallel side slots (b) one slot at the centre and (c) two side slots about the centre. The parameters S_l is the length of the slot and S_w is the width of the slot; the color yellow signifies the electrode/metal area and the orange color indicates the substrate area.

the slot. The slot affects the current path length and the electrical length of the antenna which accounts for bandwidth enhancement, frequency shifting, and PCA compactness. These interdigitated structures with slots enhance THz field strength and reduce drift time, which are compared with conventional bare-gap antenna geometries. The electric field confinement is achieved in the active area between the interdigitated electrodes due to the slots which increases the optical to THz efficiency [38]. By increasing the slot areas and reducing slot lengths, the bandwidth and electric field of PCAs are increased, respectively.

5. Theoretical modeling of IPCA

Many models like the drift-diffusion model, electronic transport model, energy balanced approximation model, finite difference time domain model to analyze the PCA performance. Based on the drift-diffusion model, the carrier density is proportional to the optical generation rate and inversely proportional to the carrier lifetime [43]. Because of the larger mass and less mobility compared with the electrons, the contribution of holes is not considered. By using the continuity equation, the time dependent carrier density can be calculated from Eq. (1):

$$\frac{\mathrm{d}n(t)}{\mathrm{d}t} = -\frac{n(t)}{\tau_c} + g(t) \tag{1}$$

where, $n(t)$ is the carrier density, τ_c is the carrier lifetime and $g(t)$ is the photocarrier generation-rate. The input optical source is considered as a gaussian laser beam. The carrier-generation rate by the laser in a PCA is given in the Eq. (2).

$$g(t) = \frac{2\eta P(t)}{\pi\omega_0^2 h\nu_{\mathrm{opt}} V} \tag{2}$$

where, η is the quantum efficiency, $P(t)$ is the optical power, ω_0 is the beam waist radius, h is Planck's constant ν_{opt} is the laser frequency, V is the active volume. The quantum efficiency η can be defined as in Eq. (3),

$$\eta = (1 - R_{\mathrm{coeff}})[1 - \exp(-\alpha T_{LT-\mathrm{GaAs}})] \tag{3}$$

where, R_{coeff} is the power reflection coefficient, α is the optical absorption coefficient, $T_{LT-\mathrm{GaAs}}$ is the skin depth. With the laser beam pulse duration (τ_l) and beam waist radius ($\omega_0\omega_0$), the optical power is described as Eq. (4).

$$P(t) = P_0\left[1 - \exp\left(\frac{-2r^2}{\omega_0^2}\right)\right]\exp\left(\frac{-2t^2}{\tau_l^2}\right) \tag{4}$$

where, P_0 is the optical peak power. The average carrier density $n_{\mathrm{avg}}(t)$ can be calculated using the Eq. (5),

$$n_{\mathrm{avg}}(t) = \frac{A.n(t)}{V_a} \tag{5}$$

In dipole PCA, the total area of the generated carriers is given as $A = L_a.W$ and the active volume is given as $V_a = L_a.W.T_{LT-\mathrm{GaAs}}$, where L_a and W are the length and width of the active area respectively. While adding interdigitated fingers in the active area at the bare gap of dipole PCA, the carriers are generated in the antenna gap between the fingers. The area of the interdigitated gap should be considered in the calculation of the IPCA structure. To know about the conduction of current across the antenna gap, the conductance can be evaluated using Eq. (6) [36],

$$G(t) = q\mu_e n(t)\left(\frac{W}{L_a}\right)T_{LT-\mathrm{GaAs}} \tag{6}$$

where, $n(t)$ is the carrier density, μ_e is the mobility of electrons, q is the charge of an electron. A restriction phenomenon of screening effect reduces THz photocurrent due to the charge polarization [44]. The $V_c(t)$ can be calculated using the radiated voltage and the screening voltage and it is written as Eq. (7).

$$V_c(t) = V_{\mathrm{bias}} - V_r(t) - V_s(t) \tag{7}$$

where, $V_c(t)$ is the gap capacitance voltage, $V_{\mathrm{bias}}(t)$ is the applied bias voltage, $V_r(t)$ is the radiating voltage, $V_s(t)$ is the screening voltage. The screening voltage can be calculated in Eq. (8) [2, 36],

$$V_s(t) = V_{\mathrm{bias}} - V_c(t) - V_r(t) \tag{8}$$

The radiated voltage can be calculated by Eq. (9),

$$V_r(t) = I_{\mathrm{total}}(t)Z_a \tag{9}$$

where, Z_a is the antenna impedance, $I_{\mathrm{total}}(t)$ is the total current. The capacitor current is given in Eq. (10),

$$I_c(t) = \frac{\pi q\mu_e^2 \Delta t A n(t) V_c^2(t)}{4S_g} \tag{10}$$

where, Δt is the time step. The total current flowing through the circuit is calculated using the Eq. (11),

$$I_{\mathrm{total}}(t) = G(t)V_c(t) + I_c(t) + \frac{V_{\mathrm{bias}}}{R_d} \tag{11}$$

where, R_d is the dark resistance. The THz photocurrent generated by the laser illumination on the semiconductor material is calculated numerically by solving the Eq. (11). The radiated THz field E_{THz} of the IPCA is directly proportional to the derivative of the terahertz photocurrent and it is calculated by Eq. (12).

$$E_{\mathrm{THz}}(t) = \frac{I_{\mathrm{total}}(t)}{V_a.G(t)} \tag{12}$$

The THz radiated power can be calculated in Eq. (13).

$$P_{\mathrm{THz}}(t) = I_{\mathrm{total}}(t)^2 Z_a \tag{13}$$

There are three types of efficiencies for a PCA, such as optical to electrical conversion efficiency η_e, THz radiation efficiency η_r and matching efficiency η_m. The overall efficiency is expressed in Eq. (14).

$$\eta_t = \frac{P_{\mathrm{THz}}(\mathrm{peak})}{P_{\mathrm{opt}}(\mathrm{peak})} \tag{14}$$

P_{opt} is the optical input power and P_{THz}(peak) is the peak terahertz radiated power. This total efficiency can be increased by increasing the η_e, as it is the ratio of electrical power to optical power.

6. Numerical modeling of IPCA

The equivalent circuit of a dipole PCA has been simulated using the PSPICE software to obtain the photocurrent. This THz photocurrent from the equivalent circuit is also compared with the fourth order of Runge-Kutta solution in MATLAB which is shown in **Figure 3a**. The photocurrent is then input to the electromagnetic model simulation of CST Microwave studio software to obtain THz wave intensity. **Figure 3b** compares THz signals based on the equivalent circuit method and Runge-Kutta method. The THz emission intensity is observed from the continuous bias, pulsed bias and spectral bandwidth.

Compared with the conventional bare-gap geometry of PCA, the interdigitated electrodes are used to increase the optical-THz conversion efficiency. The slotted IPCA and IPCA structures have been designed and simulated using the finite difference time domain simulations of CST-Microwave Studio software. The interdigitated PCA with a slot at the centre as shown in **Figure 2b** and the IPCA without slot shown in **Figure 1a** are almost overlapped to each other. The radiation from the two slots about the centre IPCA design (**Figure 2c**) has destructive interference without enhancement effect. The same interference is found from the parallel side slot IPCA design (**Figure 2a**). Compared to the other slotted IPCA designs, the centre slot IPCA structure (**Figure 2b**) provides wide bandwidth up to 1 THz. THz E-field and gain for the slotted IPCA are shown in **Figure 4a** and **b** respectively. The IPCA without slot shows a bandwidth of 0.765 THz and its gain is about 0.223 dB and 2.389 dB respectively at 4 and 5 THz. The two side slots of IPCA design increases the center frequency to 2.695 THz and results pulse width of 0.7035 ps. The two slots around the center performs the spectrum bandwidth of 1.29 Thz and the 0.782 and 3.232 dB gain respectively. The center slot IPCA performs a 0.515 THz bandwidth and 0.613 and 3.013 dB gain respectively at 4 and 5 THz. It performs high-intensity THz wave generation [37]. The tip-to-tip IPCA shown in **Figure 1c** has the electric-field of about 4.5×10^6 V/m. The bow-tie antenna coupled with IPCA (**Figure 1b**) can enhance the THz generation.

The time-dependent average carrier density of IPCA is plotted via Eq. (5) and MATLAB software in **Figure 5a**. The carriers are largely generated in a large active area gap based on Eq. (1). The carrier density is high in a dipole PCA, comparing to the designed IPCA as it has a large gap between the anode and cathode. The n_{avg} of a

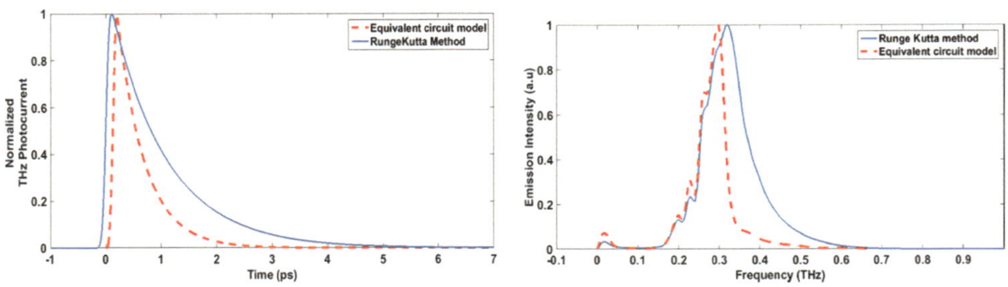

(a) (b)

Figure 3.
(a) Normalized THz photocurrent (b) intensity of THz signal (reprinted from [45] with permission of Springer publishing).

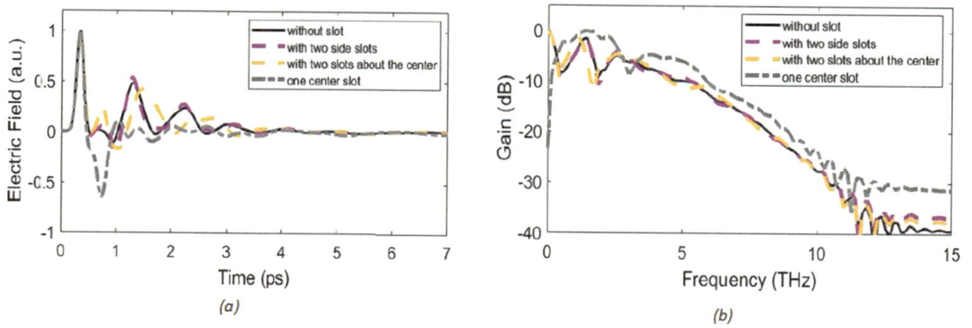

Figure 4.
(a) Simulated E-field for the slotted IPCA and (b) gain for the slotted IPCA (reprinted from [21], with permission of IEEE publishing).

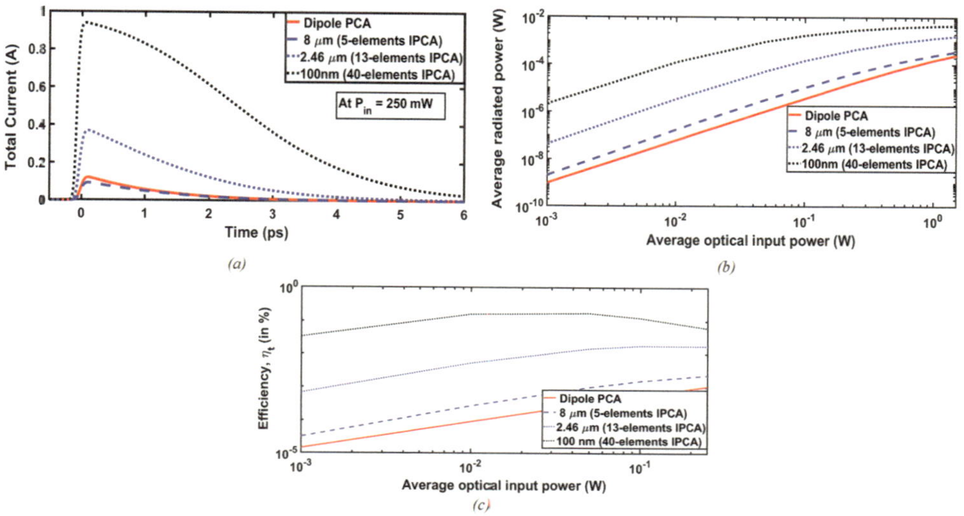

Figure 5.
Simulated results of (a) Total current of dipole and IPCA (b) radiated power of dipole and IPCA and (c) optical-THz conversion efficiency of dipole and IPCA.

dipole PCA is about 4.89×10^{23} m^{-3}. For the 100 nm IPCA is 1.02×10^{23} m^{-3}. Due to a larger gap, the screening effect become obvious to reduce THz intensity generation. **Figure 5b** shows the current conduction across the IPCA gap based on Eq. (6). The interdigitated gap conductance $G(t)$ depends on the carrier mobility, laser input power, carrier lifetime, antenna geometry and laser pulse duration. When the number of metal teeth structures increases in an active area, the conductance increases with the electric field amplitude based on Eq. (12). The IPCA teeth width of 100 nm possess conductance of 0.15048 ℧ whereas the conventional dipole PCA possess 0.000173 ℧ with 250 mW input optical power. To calculate THz radiation power of an IPCA, the antenna radiation resistance is required based on Eq. (13). Hence, the IPCA can be designed using the Finite Element Method (FEM) within Ansys High-Frequency Structure Simulator (HFSS) package. Among the real and imaginary part resistance of an antenna, only the real part resistance is dependent on THz frequency. All the IPCA designs radiates maximum power at 0.4 THz and hence the real part ohmic resistance is considered at that particular frequency which is shown in **Figure 5c**.

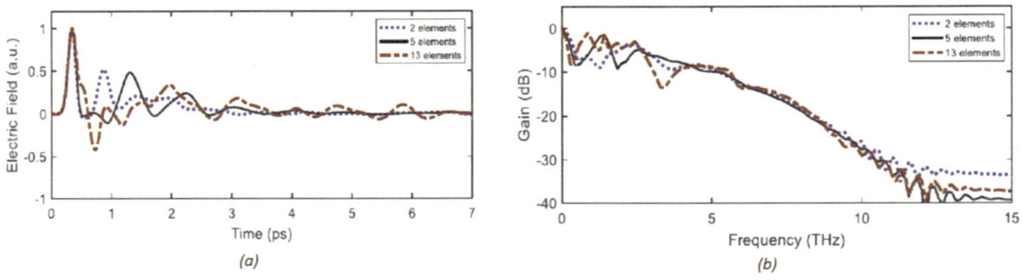

Figure 6.
(a) Simulated E-field for the IPCA and (b) gain for the different elements of IPCA (Reprinted from [21], with permission of IEEE publishing).

The IPCA total current, THz radiation power and Opt-THz conversion efficiency of dipole PCA and IPCA structures are plotted in the **Figure 6** based on Eqs. (1)–(14). The antenna gap widths are 1μm for all IPCA designs. The amplitude of current gets increased by high carrier generation. In PCA, the antenna radiates toward the substrate. THz radiation power and efficiency are plotted based on Eqs. (13) and (14) respectively. Increasing the input pump power leads to high radiation power. Even with low input power of 250 mW, the IPCA of width 100 nm (40-elements) radiates up to 2.962 mW. The total current for the 40-elements IPCA is about 0.939 A whereas 0.12 A for the dipole PCA. The total efficiency for the dipole PCA is about 0.001% and that of the 40-elements IPCA is about 0.06%. Comparing to the dipole PCA, an IPCA efficiency as high as 60 orders of magnitude.

By increasing the interdigitated elements in the active area of an IPCA, the enhanced electric field is obtained, which is shown in **Figure 7a**. **Figure 7b** depicts the corresponding spectra which clearly show the gain fluctuation for 2, 5 and 13 interdigitated elements. The 2, 5 and 13-elements possess peak intensities of 24.86, 26.42 and 27.66 W/cm^2 respectively. Hence, designing the antenna with the various numbers of interdigitated elements can control THz radiation spectra.

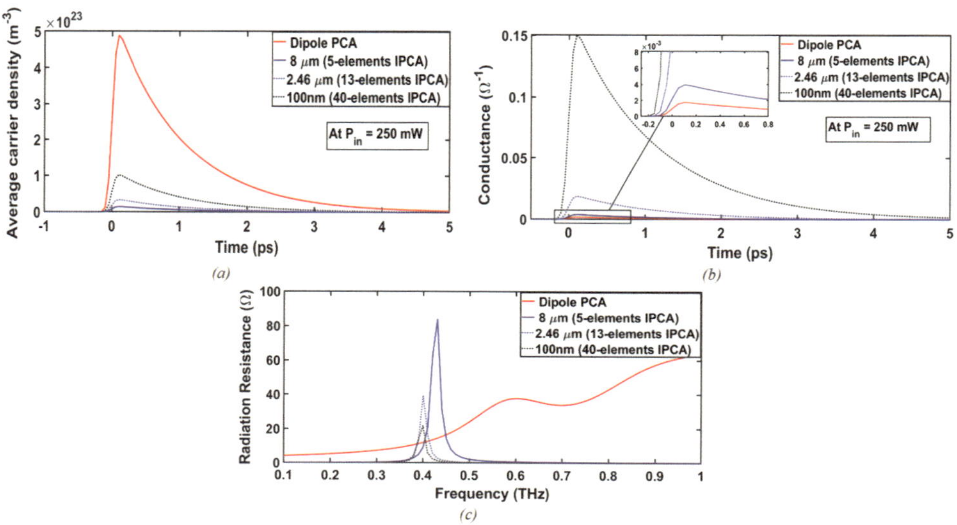

Figure 7.
Simulated results (a) average carrier density of dipole and IPCA (b) gap conductance of dipole and IPCA and (c) antenna resistance of dipole and IPCA.

7. Experimental results of IPCA

The experimental results of IPCA antennas are presented in this section. The micrograph of an IPCA device and its cross sectional view are given in **Figure 8a** and **b** respectively. The fabrication process of this device involves the following steps. First of all, the electrodes are deposited on top of a lattice-matched layer system consisting of 150 nm SI-GaAs, a 100 nm AlAs layer, followed by a 1.3 μm layer of LT-GaAs. The LT-GaAs is grown using molecular beam epitaxy on a SI-GaAs substrate. The photolithography technique has been used to pattern the electrode using mask and photoresist. The metal along with AlAs layer has been etched out in a HF solution resulting the gap in interdigitated electrode structure, shown in **Figure 8b**. The IPCA array chip is then transferred to an optically and terahertz transparent sapphire substrate of 500 μm thickness.

To measure the THz radiation from IPCA single element and IPCA array, the device is placed in a standard confocal terahertz-time-domain spectroscopy system (THz-TDS) and pumped by the optical beam from an ultrafast Ti:sapphire laser, operating at 780 nm having a repetition rate of 76 MHz with a pulse width of 100 fs. The IPCA has been placed normally to the femtosecond laser beam and the terahertz radiation exits the substrate (sapphire) side along with the residual pump beam. For the generation of THz signal from the IPCA, the substrate lens is not used. However, a silicon substrate lens is used for THz detection. The typical bias voltage of 0–40 V amplitude square wave is used for biasing. For the detection system, the same IPCA is used with 3 mm diameter silicon substrate lens to focus the terahertz radiation onto the photoconductive gap. The normalized electric field for the IPCA is obtained from THz-TDS and the corresponding frequency spectra are depicted in **Figure 9**. The figure shows around 30% increase in THz amplitude between the single and array IPCA. In order to identify the reason for increasing THz radiation, the optical excitation area is measured (**Figure 9**). Due to higher amount of radiation, a strongly directed THz beam (**Figure 10**) produces a large

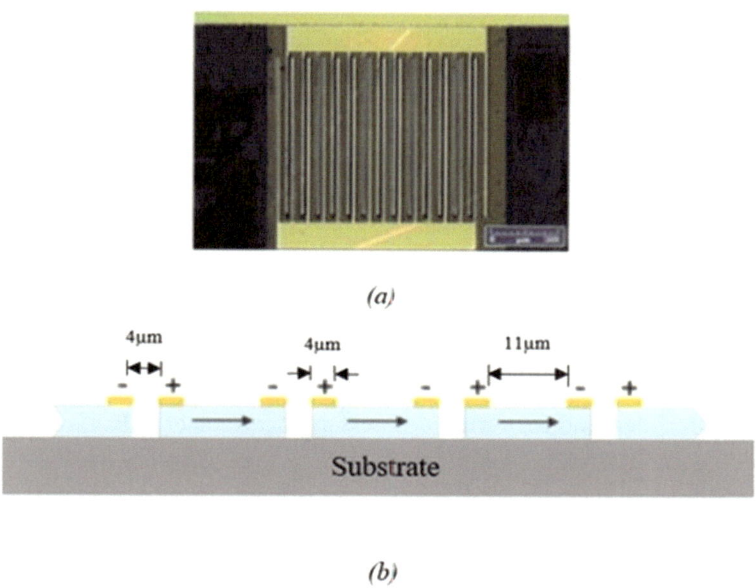

Figure 8.
(a) Micrograph of terahertz IPCA device. (b) Cross-sectional view of terahertz IPCA array (reprinted from [46] with permission of Applied Physics Letters publishing).

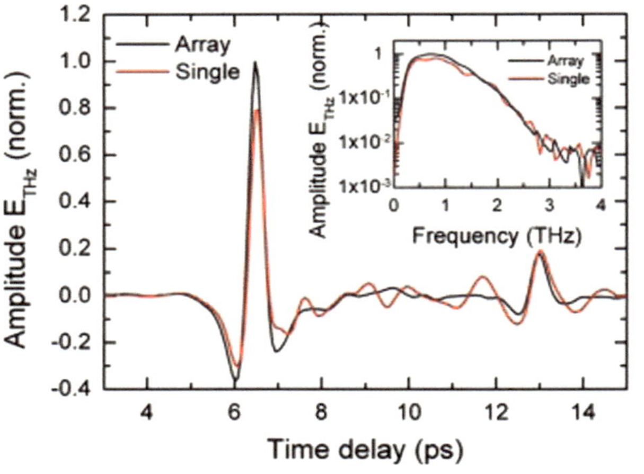

Figure 9.
Time-resolved scans of terahertz pulse for IPCA array emitter (dark line) and single element IPCA (light line) and their respective spectra in inlay (reprinted from [46] with permission of Applied Physics Letters publishing).

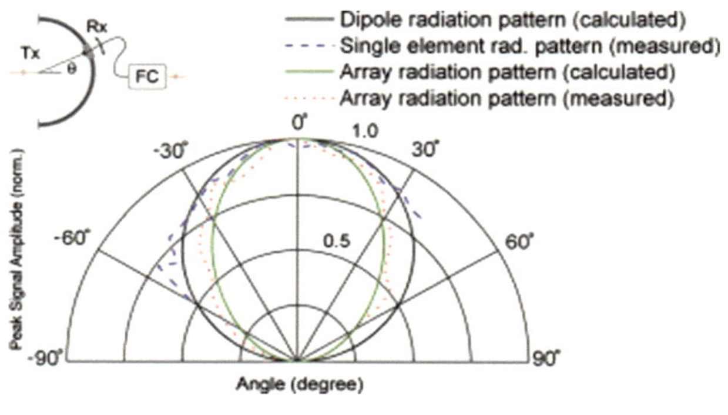

Figure 10.
(Color online) Calculated radiation pattern for dipole (black solid) line and measured single element IPCA radiation pattern (dashed line). IPCA array radiation pattern calculated based on array theory (light solid line) and measurement (dotted). Inlay: experimental setup with fiber coupled and dispersion compensated (FC) receiver for radiation pattern measurement (reprinted from [46] with permission of Applied Physics Letters publishing).

THz wave amplitude at the detector which is collected using the parabolic mirrors in the measurement setup.

8. Conclusion

The enormous demand for future wireless applications of THz waves motivates the researchers to target on the development of THz sources and detectors. Among all the conventional PCAs, IPCAs improve THz generation and detection. The analytical calculations of IPCA helps us to understand capacitive behavior of the gap between the electrodes. From the above discussed results, it has been observed that the photocurrent and optical to THz conversion efficiency are enhanced by the IPCA design compared with the dipole PCA. THz performance can be further improved by introducing the non-linear effect in the IPCA active area or by the plasmonic features.

Author details

Shyamal Mondal[1]*, Nisha Flora Boby Edwin[2] and Vaisshale Rathinasamy[2]

1 Department of Applied Physics, Defense Institute of Advanced Technology, Pune, India

2 Department of Electronics and Communication Engineering, SRM Institute of Science and Technology, Chennai, India

*Address all correspondence to: shyamal.kgec@gmail.com

References

[1] Tonouchi M. Cutting-edge terahertz technology. Nature Photonics. 2007; **1**(2):97-105

[2] Khiabani N, Huang Y, Shen YC, Boyes S. Theoretical modeling of a photoconductive antenna in a terahertz pulsed system. IEEE Transactions on Antennas and Propagation. 2013;**61**(4): 1538-1546

[3] Yardimici NT, Yang SH, Jarrahi M. High power pulsed terahertz radiation from large area plasmonic photoconductive emitters. In: 2015 40th International Conference on Infrared, Millimeter, and Terahertz waves (IRMMW-THz). Conference and proceedings: IEEE; 2015. pp. 1-2

[4] Collier CM, Stirling TJ, Hristovski IR, Krupa JD, Holzman JF. Photoconductive terahertz generation from textured semiconductor materials. Scientific Reports. 2016;**6**(1):1-10

[5] Headley C, Fu L, Parkinson P, Xu X, Lloyd-Hughes J, Jagadish C, et al. Improved performance of GaAs-based terahertz emitters via surface passivation and silicon nitride encapsulation. IEEE Journal of Selected Topics in Quantum Electronics. 2010;**17**(1):17-21

[6] Mitrofanov O, Brener I, Luk TS, Reno JL. Photoconductive terahertz near-field detector with a hybrid nanoantenna array cavity. ACS Photonics. 2015;**2**(12):1763-1768

[7] Yang SH, Hashemi MR, Berry CW, Jarrahi M. 7.5% optical-to-terahertz conversion efficiency offered by photoconductive emitters with three-dimensional plasmonic contact electrodes. IEEE Transactions on Terahertz Science and Technology. 2014;**4**(5):575-581

[8] Bashirpour M, Kolahdouz M, Neshat M. Enhancement of optical absorption in LT-GaAs by double layer nanoplasmonic array in photoconductive antenna. Vacuum. 2017;**146**:430-436

[9] Ghorbani S, Bashirpour M, Poursafar J, Kolahdouz M, Neshat M, Valinejad A. Thin film tandem nanoplasmonic photoconductive antenna for high performance terahertz detection. Superlattices and Microstructures. 2018;**120**:598-604

[10] Bashirpour M, Ghorbani S, Forouzmehr M, Kolahdouz M, Neshat M. Optical absorption enhancement in LTG-GaAs for efficiency improvement of THz photoconductive antennas. In: 2016 Fourth International Conference on Millimeter-Wave and Terahertz Technologies (MMWaTT). Conference and proceedings: IEEE; 2016. pp. 14-16

[11] Lepeshov S, Gorodetsky A, Krasnok A, Toropov N, Vartanyan TA, Belov P, et al. Boosting terahertz photoconductive antenna performance with optimised plasmonic nanostructures. Scientific Reports. 2018; **8**(1):1-7

[12] Yardimci NT, Jarrahi M. High sensitivity terahertz detection through large-area plasmonic nano-antenna arrays. Scientific Reports. 2017;**7**(1):1-8

[13] Akbar F, Kolahdouz M, Larimian S, Radfar B, Radamson H. Graphene synthesis, characterization and its applications in nanophotonics, nanoelectronics, and nanosensing. Journal of Materials Science: Materials in Electronics. 2015;**26**(7):4347-4379

[14] Nguyen TK, Park I. Effects of antenna design parameters on the characteristics of a terahertz coplanar stripline dipole antenna. Progress in Electromagnetics Research. 2013;**28**: 129-143

[15] Zhu N, Ziolkowski RW. Photoconductive THz antenna designs with high radiation efficiency, high directivity, and high aperture efficiency. IEEE Transactions on Terahertz Science and Technology. 2013;**3**(6):721-730

[16] Alias NB, Awang AH. Physical parametric analysis of Terahertz Photoconductive bow-tie dipole Antenna on frequency and radiation pattern using electromagnetic simulation tools. In: 2016 IEEE Asia-Pacific Conference on Applied Electromagnetics (APACE). IEEE; 2016. pp. 263-267

[17] Khiabani N. Modelling, Design and Characterisation of Terahertz Photoconductive Antennas. Ireland, UK: The University of Liverpool; 2013

[18] Zhang J, Tuo M, Liang M, Ng WR, Gehm ME, Xin H. Terahertz radiation of a butterfly-shaped photoconductive antenna. Hongwai yu Jiguang Gongcheng/Infrared and Laser Engineering. 2019;**48**(4):0402001

[19] Bashirpour M, Forouzmehr M, Hosseininejad SE, Kolahdouz M, Neshat M. Improvement of terahertz photoconductive antenna using optical antenna array of ZnO nanorods. Scientific Reports. 2019;**9**(1):1-8

[20] Mondal S, Rathinasamy V, Kapoor S, Mukherjee S, Rao TR. Interdigitated photoconductive antenna design and analysis for terahertz wireless applications. In: 2020 IEEE 3rd 5G World Forum (5GWF). IEEE; 2020. pp. 484-487

[21] Tucek J, Gallagher D, Kreischer K, Mihailovich R. A compact, high power, 0.65 THz source. In: 2008 IEEE International Vacuum Electronics Conference. Conference and proceeding: IEEE; 2008. pp. 16-17

[22] Gregory IS, Baker C, Tribe WR, Bradley IV, Evans MJ, Linfield EH, et al. Optimization of photomixers and antennas for continuous-wave terahertz emission. IEEE Journal of Quantum Electronics. 2005;**41**(5):717-728

[23] Duffy SM, Verghese S, McIntosh A, Jackson A, Gossard A, Matsuura S. Accurate modeling of dual dipole and slot elements used with photomixers for coherent terahertz output power. IEEE Transactions on Microwave Theory and Techniques. 2001;**49**(6):1032-1038

[24] Ferguson B, Zhang XC. Materials for terahertz science and technology. Nature Materials. 2002;**1**(1):26-33

[25] Cai Y, Brener I, Lopata J, Wynn J, Pfeiffer L, Federici J. Design and performance of singular electric field terahertz photoconducting antennas. Applied Physics Letters. 1997;**71**(15): 2076-2078

[26] Auston DH. Picosecond optoelectronic switching and gating in silicon. Applied Physics Letters. 1975; **26**(3):101-103

[27] Pozar D. Chebyshev multisection matching transformers. In: Microwave Engineering. 4th ed. Hoboken, NJ, USA: John Wiley & Sons, Inc; 2012. pp. 256-261

[28] Singh D, Kalialakis C, Gardner P, Hall PS. Small H-shaped antennas for MMIC applications. IEEE Transactions on Antennas and Propagation. 2000; **48**(7):1134-1141

[29] Llatser I, Kremers C, Cabellos-Aparicio A, Jornet JM, Alarcón E, Chigrin DN. Graphene-based nano-patch antenna for terahertz radiation. Photonics and Nanostructures-Fundamentals and Applications. 2012; **10**(4):353-358

[30] Deng H, Xiong Z, Qu W, Wu Z, Liu Q, Chen L, et al. The impact of structural parameters of split-ring resonators on the terahertz radiation

characteristics of micro-structured photoconductive antennas: A simulation study. IEEE Photonics Journal. 2020;**12**(3): 1-13

[31] Tani M, Matsuura S, Sakai K, Nakashima SI. Emission characteristics of photoconductive antennas based on low-temperature-grown GaAs and semi-insulating GaAs. Applied Optics. 1997;**36**(30):7853-7859

[32] Mikulics M, Michael E, Schieder R, Stutzki J, Güsten R, Marso M, et al. Traveling-wave photomixer with recessed interdigitated contacts on low-temperature-grown GaAs. Applied Physics Letters. 2006;**88**(4):041118

[33] Vieweg N, Mikulics M, Scheller M, Ezdi K, Wilk R, Hübers HW, et al. Impact of the contact metallization on the performance of photoconductive THz antennas. Optics Express. 2008; **16**(24):19695-19705

[34] Matthäus G, Nolte S, Hohmuth R, Voitsch M, Richter W, Pradarutti B, et al. Microlens coupled interdigital photoconductive switch. Applied Physics Letters. 2008;**93**(9):091110

[35] Ropagnol X, Morandotti R, Ozaki T, Reid M. THz pulse shaping and improved optical-to-THz conversion efficiency using a binary phase mask. Optics Letters. 2011;**36**(14):2662-2664

[36] Prajapati J, Bharadwaj M, Chatterjee A, Bhattacharjee R. Circuit modeling and performance analysis of photoconductive antenna. Optics Communications. 2017;**394**:69-79

[37] Rathinasamy V, Kapoor S, Rout A, Rao TR, Mondal S. Interdigitated-slot photoconductive antenna for terahertz applications. In: IEEE Indian Conference on Antennas and Propogation (InCAP). IEEE; 2019, 2019. pp. 1-3

[38] Rathinasamy V, Thipparaju RR, Edwin NFB, Mondal S. Interdigitated photoconductive terahertz antenna for future wireless communications. Microwave and Optical Technology Letters. 2021

[39] Boby ENF, Rathinasamy V, Rao TR, Mondal S. Parametric analysis of inter-combed photoconductive antenna for terahertz communication. In: 2021 International Conference on Communication information and Computing Technology (ICCICT). Conference and proceedings: IEEE; 2021. pp. 1-4

[40] Berry CW, Wang N, Hashemi MR, Unlu M, Jarrahi M. Significant performance enhancement in photoconductive terahertz optoelectronics by incorporating plasmonic contact electrodes. Nature Communications. 2013;**4**(1):1-10

[41] Tanoto H, Teng J, Wu Q, Sun M, Chen Z, Maier S, et al. Greatly enhanced continuous-wave terahertz emission by nano-electrodes in a photoconductive photomixer. Nature Photonics. 2012; **6**(2):121-126

[42] Seo M, Park H, Koo S, Park D, Kang J, Suwal O, et al. Terahertz field enhancement by a metallic nano slit operating beyond the skin-depth limit. Nature Photonics. 2009;**3**(3):152-156

[43] Brown E, Smith F, McIntosh K. Coherent millimeter-wave generation by heterodyne conversion in low-temperature-grown GaAs photoconductors. Journal of Applied Physics. 1993;**73**(3):1480-1484

[44] El-Ghazaly SM, Joshi RP, Grondin RO. Electromagnetic and transport considerations in subpicosecond photoconductive switch modeling. IEEE Transactions on Microwave Theory and Techniques. 1990;**38**(5):629-637

[45] Nissiyah GJ, Madhan MG. A simple equivalent circuit model of photoconductive dipole antenna for the

study of terahertz intensity modulation. Optical and Quantum Electronics. 2021; **53**(4):1-22

[46] Awad M, Nagel M, Kurz H, Herfort J, Ploog K. Characterization of low temperature GaAs antenna array terahertz emitters. Applied Physics Letters. 2007;**91**(18):181124

Chapter 10

Terahertz Conductivity of Nanoscale Materials and Systems

Rahul Goyal and Akash Tiwari

Abstract

The history of RF technology can provide human beings a powerful lesson that the infrastructure of modern-day wireless communication depends on the complexity and configurability of silicon-based solid-state devices and integrated circuits. The field of THz technology is undergoing a developmental revolution which is at an inflection point and will bridge the 'technology' and 'application' gap in meaningful ways. This quantitative progress is a result of continuous and concerted efforts in a wide range of areas including solid-state devices, 2D materials, heterogeneous integration, nanofabrication and system packaging. In this chapter, the innovative theoretical approaches that have enabled significant advancement in the field of system-level THz technology are discussed. The focus is kept on the formulation of terahertz conductivity which plays a critical role in the modeling of devices that integrate technologies across electronics and photonics. Further, the findings build on coupling a probe pulse of terahertz illumination into the photoexcited region of amorphous silicon are presented and discussed in detail. Terahertz light has a higher penetration depth for opaque semiconductor materials which provides an accurate method to measure the conductivity of novel materials for the construction of efficient solar cells. This paves the way for the possibility to develop energy systems can address the need for reconfigurability, adaptability and scalability beyond the classical metrics.

Keywords: terahertz, solidstate, electronics, semiconductors, quantum

1. Introduction

Terahertz radiation generally accounts for the photons of energy ranging from 414 μeV to 41.4 eV. This energy spectrum corresponds to the frequencies of 0.1–10 THz in the electromagnetic spectrum. In the past, it has been referred to as the 'terahertz gap' primarily due to the inefficient methods for generation and detection of waves in the frequency range [1, 2]. The major reason for the complexity in the designing of efficient THz systems was the presence of background noise sources in incoherent light (room temperature is 25 meV, or 6 THz) [2–4]. Over the past few decades, due to the advancement in the nanoscale fabrication, there has been a significant surge of progress in enabling integrated, compact and efficient chip-scale solid-state semiconductor technology that can operate at room temperature and can be manufactured at a low cost, exploiting economies of scale. Considerable work has been directed towards miniaturized technologies demonstrated with quantum-cascade lasers [5], microbolometers [6], nanowires [7], novel plasmonic nanostructures [8], metamaterials [9] and ultrafast conductive semiconductor materials [10]. This progress has resulted a major step towards a more holistic approach for realizing the development of THz systems for applications in communication, spectroscopy and hyperspectral imaging [11].

Interaction of the terahertz radiation with matter provides a vital low-energy probe to the electronic nature of doped semiconductors by generating carriers optically with a fraction of laser beam [12, 13], this technique is known as Optical Pump Terahertz Probe Spectroscopy (OPTPS). This technique has the benefit of allowing the carrier density to be readily controllable, and also allows the photoconductivity to be measured on picosecond time scales after photoexcitation. A detailed description of the OPTPS is presented by the authors of paper; [14], and further, the authors discussed the terahertz spectroscopy method to measure transient variation in the photoconductivity of bulk and nanostructured semiconductors. There are a number of terahertz spectroscopic techniques and applications which have been presented and discussed from various scientists around the world. The studies based on the terahertz spectroscopy of polymer-fullerene based hetrojunctions are reviewed by the authors of [15] and the utilization of near-field terahertz for ultrafast imaging is explained in depth by the authors of [16].

On the electronics front for the development of terahertz-based devices and integrated circuits (ICs), researchers across various disciplines have converged to address the technological challenges in the field. Scientists have made numerous efforts in engineering the fabrication of novel nanomaterials, and most importantly, in the development of theoretical framework to optimize the design parameters of the devices and ICs. The main result of this effort is reflected in the ability of various technologies to generate terahertz signals with sufficient power levels. Particularly in frequencies beyond 1 THz, the semiconductor technologies based on III–V group elements such as InP heterojunction bipolar transistors (HBTs), high electron mobility transistors (HEMTs) and GaAs-based Schottky diodes are now capable of generating power levels almost two to five times higher than a decade ago [17, 18]. Silicon-based integrated technology, providing a platform for massive integration has demonstrated complex phased array, imaging and communication systems with output power reaching up to the 100 μW at the terahertz frequency range [19, 20].

Semiconductor device modeling represents a physics-based analytical modeling approach to predict device operations at specific conditions such as applied bias (voltages and currents), environment (temperature and noise) and physical characteristics (geometry and doping levels). The fundamental understanding of the charge carrier dynamics plays a crucial role in the formulation of numerical models by implementing mathematically fitted conductivity equations. Numerical models are primarily used as virtual environments for device optimization to directly estimate the conductivity providing insight into the nature of charge carriers, their mobility and its dependence on time. In this chapter, the focus is on the conductivity at terahertz frequencies of bulk and nano-structured solid state materials. A variety of theoretical formalisms that describe the terahertz conductivity of bulk, mesoscopic and nanoscale materials are explained in the chapter (Section 2). The validity and limitations of the respective formulations are discussed to highlight the boundary of implementation for accurate calculations. The chapter starts with the definition of the complex conductivity (Section 3) and then the experimental method (Section 4) to obtain complex conductivity is described to develop the understanding of practical implementation. Finally, the effectiveness of surface passivation using hydrogenated amorphous silicon is quantified through time resolved terahertz spectroscopy and the results of the measurements are presented in the form of change in photoconductivity upon excitation from terahertz electric

DOI: http://dx.doi.org/10.5772/intechopen.104797

field (Section 6). Further, in the section, the results of Fourier transform infrared spectroscopy are discussed to comment upon the nature of bond formation in the amorphous silicon passivation layer.

The complex conductivity (σ) can be defined as a mathematical term that relates the current density vector ($\vec{J}$) to the electric field vector ($\vec{E}$). This relation is similar to the way that current (I) and voltage (V) in an ac circuit are related by the complex conductance (G), the stated analogy can be understood by the following set of equations,

$$I = G \times V \tag{1}$$

$$G = \sigma \times \frac{A}{L} \tag{2}$$

$$V = E \times L \tag{3}$$

$$J = \frac{I}{A} = \frac{\sigma \times \frac{A}{L} \times E \times L}{A} \tag{4}$$

where L is the length of the conductor and A is the cross-sectional area of the conductor. Hence in vector form,

$$\vec{J} = \sigma \vec{E}. \tag{5}$$

The response of a material to an electromagnetic wave is generally described by the dielectric constant of the material. Fundamentally, the refractive index of a material quantifies the combined response of a material to a electric and magnetic field which is given by the following equation,

$$\tilde{n} = n + i\kappa = \sqrt{\varepsilon\mu} \tag{6}$$

where the dielectric function or electrical permittivity (ε) describes the ability of an electric field to penetrate the material medium and the magnetic permeability (μ) quantifies magnetic response of the medium. The dielectric function is generally complex with the significantly higher imaginary part (ε_2) as compared to the real part (ε_1) in spectral regions with dominant absorption.

$$\varepsilon = \varepsilon_1(\omega) + i\varepsilon_2(\omega) = |\varepsilon|e^{-i\delta} = |\varepsilon|(\cos\delta - i\sin\delta) \tag{7}$$

where ε_1 and ε_2 is the real and imaginary part of the permittivity and δ is the loss angle. It is often convenient to disentangle the contribution to ε from bound modes (lattice vibrations and core electrons) and the component arising from free charges. In the THz and mid-IR frequency range, lattice vibrations contribute via transverse optical phonon absorption (in polar materials), while at optical frequencies interband transitions alter ε. The lattice component can be derived from the Maxwell's equations and can be mathematically written in the following way,

$$\varepsilon(\omega) = \varepsilon_L(\omega) + \frac{i\sigma(\omega)}{\omega\varepsilon_0} \tag{8}$$

where ω is the angular frequency of the electromagnetic wave, σ is the electrical conductivity, ε_0 is the electrical permittivity of free space and $\varepsilon_L(\omega)$ is the component of ε from bound modes (lattice vibrations and core electrons). The electrical conductivity (σ) of mobile charges is also defined complex in nature and hence can be written as,

$$\sigma(\omega) = \sigma_1(\omega) + i\sigma_2(\omega) \tag{9}$$

Generally, a valid assumption for the equations derived from Maxwell's equation is the excitation from a conventional monochromatic plane wave propagating in the $\hat{k}$ direction, with a time dependent electrical field, which is defined as,

$$E(t) = E_0 e^{i(k.r-\omega t)} \tag{10}$$

where E_0 represents the maximum electric field strength, r is the three-dimensional position vector of a point, and $k \cdot r$ denotes the projection of the direction vector $\hat{k}$ along the position vector r.

2. Models for conductance at terahertz frequency

In this chapter, various models for the terahertz conductivity are compared and outlined in sequential manner, initially the basic fundamental physics of charge transport is described to develop the understanding of terahertz conductivity in a homogeneous media. The chapter starts with the description of Drude-Lorentz model which is applicable in classical domain of charge transport and further discussed about the limitation in the applicability of the model. Further, the chapter provides the discussion of the models that functions effectively at the quantum mechanical domain and elucidates the mathematical expressions for the terahertz conductivity. The relationship between the length scale and the applicability of various conductivity models is demonstrated with the help of a schematics shown in **Figure 1**, it clearly separates the quantum and classical conductivity models based on the length scale of electrons mean free path in the semiconductor. The schematics also shows the conductivity models namely Drude-Smith and Plasmon model which consider the interfacial scattering of electrons into account for the sub-micrometer length scale of the semiconductor material.

3. Complex conductivity

A classical circuit theory forms a simple analog to examine the complex conductivity $\sigma = \sigma_1 + i\sigma_2$. A voltage wave of the mathematical form $V = V_0 e^{-i\omega t}$ generates

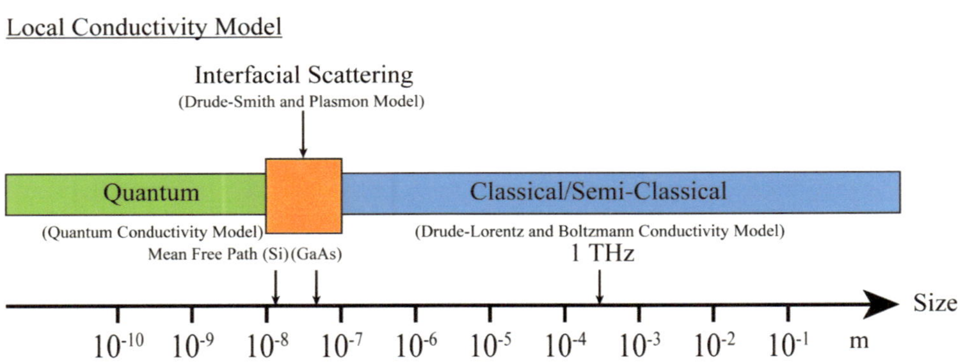

Figure 1.
The relevant terahertz conductivity models of blends and micro/nano-materials at different length scales.

a complex impedance $Z = Z_R + i|(Z_L + Z_C) = R + i(-\omega L + 1/\omega C)$ where R, L and C are resistance, inductance and capacitance of the circuit respectively. A non-polarisable material provides a good model system to theoretically understand the functioning of a bulk semiconductor with a single charge carrier, in which the current lags behind the applied voltage (a resistor R in series with an inductor L). It is known that conductance is defined as an inverse of resistance, hence it can be written that the complex conductivity (σ) is proportional to the inverse of complex impedance (Z) and therefore, mathematically it can be expressed as, $\sigma \propto 1/Z \propto R + i\omega L$. Similarly, the charge transport in a polarisable material can be examined by including a capacitor C in the equivalent circuit. A circuit consisting of resistor and capacitor in series configuration, the complex impedance is given by $Z = R + i/\omega C$ and the complex conductivity is proportional to $\sigma \propto 1/Z \propto R - i/\omega C$. It should be noted that in the case of non-polarisable material the imaginary part of complex conductivity is positive but for polarisable material it has a negative sign. This property plays a very important role in the mechanics of charge transportation for polarisable medium.

3.1 Drude-Lorentz model

Drude-Lorentz model provides the simplest approach to mathematically interpret the frequency-dependent conductivity of metals and semiconductors. This model considers the charge carriers i.e. electrons and holes transportation as a non-interacting gas plasma. It is assumed in the model that the charge carriers undergo random collisions at a rate $\Gamma = 1/\tau$ where τ denotes the relaxation time between collisions and Γ is assumed to be independent of energy in the system. According to the Drude-Lorentz model, the displacement x of the charge carrier with effective mass m^* and charge q from its equilibrium position is given by,

$$\frac{d^2x}{dt^2} + \Gamma\frac{dx}{dt} = \frac{qE_{\text{local}}}{m^*} \tag{11}$$

where at normal incidence the local electric field E_{local} of the terahertz pulse is related to the incident electric field E_{THz} by the following equation, $E_{\text{local}} = 2E_{\text{THz}}/(n+1)$. The term $qE_{\text{local}}\tau/m^*$ represents the velocity by which the charge carriers drift in the presence of constant applied field. It can be assumed from the linear nature of the differential equation that if the incident electric field oscillates at angular frequency ω and has the $E_{\text{THz}} = E_0e^{-i\omega t}$ mathematical form, then the displacement x has the solution of an identical time variation i.e. $x = x_0e^{-i\omega t}$. The value of x_0 can be yielded by substituting the terms of E and x in Eq. (11) which represents the mean oscillating amplitude of the charge carrier. The dielectric function ε of the medium can be determined from the definition of the polarization from $P = \chi\varepsilon_0E = Nqx$, and then the conductivity as a function of frequency can be expressed from the following equation,

$$\sigma(\omega) = \frac{Nq^2}{m^*}\frac{\tau}{1 - i\omega\tau}. \tag{12}$$

It should be noted form the equation that the real part (σ_1) of the complex conductivity has a maximum at the zero frequency and similarly the imaginary part has a maximum at the frequency $\omega = 1/\tau$. Intraband conductivity of some metals and semiconductors are adequately modeled by the Drude-Lorentz conductivity, however there are some variables which are considered constants for mathematical simplicity, such as electron scattering rate which is dependent on energy, has been considered constant in the Drude-Lorentz model. Most importantly, it is assumed in the model that the material is uniform over the length scale that is traversed by electrons during their motion which often significantly varies in the experimental observations [21]. These limitations are the fundamental reason for the Drude-Lorentz model to be invalid for two-dimensional materials, having examined these limitations it is important to consider alternative terahertz conductivity models to establish more accurate simulation environment.

3.2 Plasmon model

The plasmon model is a straightforward extension to the Drude-Lorentz model, in this model a restoring force with an external electromagnetic wave derives the motion of electrons which is governed by the equation of motion of a damped, driven simple harmonic oscillator [22]. The restoring force can be provided by electromagnetic or for instance electrostatic, by a surface depletion layer or accumulation layer, which is often the case when calculating for the charge transportation at the interface of semiconductors. Hence, the plasmon model is mostly applicable to model the terahertz conductivity of semiconductor nanomaterials. A mathematical term $\omega_0^2 x$ is added to the left-hand side of Eq. (11) which represents the applied restoring force and where ω_0 is the angular frequency of the oscillatory response.

$$\frac{d^2x}{dt^2} + \Gamma\frac{dx}{dt} + \omega_0^2 x = \frac{qE_{\text{local}}}{m^*} \tag{13}$$

A larger restoring force requires a greater value of ω_0. The complex conductivity in the model is given by the equation,

$$\sigma(\omega) = \frac{N \times e^2}{m^*} \frac{\tau}{\left(1 - \left(i\tau\left(\omega - (\omega_0)^2/\omega\right)\right.\right.} \tag{14}$$

The oscillation frequency (ω_0) can be mathematically linked to plasmon frequency (ω_p) for some particular geometries by the following equations,

$$\omega_p = \sqrt{\frac{N \times e^2}{m^* \varepsilon_\infty \varepsilon_0}} \tag{15}$$

and

$$\omega_0^2 = f \times \omega_p^2 \tag{16}$$

The equations are valid for the assumption that the charges are located on the surface of a small spherical particle which has the geometric factor $f = 1/3$ [22]. while for cylindrical wires $f = 0$ when the electric field E_{THz} is axial and the wave vector q_{THz} is radial, $f = 1/2$ when E_{THz} and q_{THz} are radial, and $f = 1/3$ for radial E_{THz} and axial q_{THz} [23, 24]. There has been many practical implementation of the classical conductivity expression such as it has been applied in temperature-dependent terahertz time domain spectroscopy studies of doped silicon, however there are several reasons why the classical relaxation-effect model is not sufficient for a quantitative account of experimental observation, even when interband transitions can be safely ignored or calculated separately:

1. Conduction bands with normal metals can be significantly different from the ideal spherical energy band (that assumes an effective electron mass, m).

2. Scattering relaxation time is assumed to be independent of energy $E(k)$, where k = wave vector of the electron, even though it depends on both energy and position. For this reason, τ is considered to be a semi-empirical parameter.

3. A local-response regime is assumed.

The above reasons highlight the importance to achieve even better modeling accuracy and therefore it is necessary to move towards the more complicated semiclassical treatment of terahertz conductivity of materials.

3.3 Phenomenological conductivity model

In the Drude-Lorentz model, it is assumed that the charge carriers are propagating in the bulk of the material which means that the carrier displacement is less than the largest dimension L of the material. Thus, it can be argued that when an electron or hole's displacement under the electric field E_{THz} becomes comparable to the dimension L of the material then the conductive response of the material will differ significantly from that of the bulk material. Mathematical, this can be understand by the criterion of mean free path such that the electron or hole mean free path l should be greater than L, i.e. $l = v\tau > L$, where the carrier velocity is often equated with the thermal velocity which is given by, $v = \sqrt{\frac{3k_BT}{m^*}}$. The carrier velocity defines the This relation can be demonstrated by some examples; firstly, for room temperature electrons close to the Γ-valley minimum of GaAs with the values of $v \sim 5 \times 10^5$ m/s and $\tau \sim 100$ fs, this yields the mean free path (l) of 50 nm and thus, the dimension L of the material should be less than 50 nm. Secondly, for silicon $m^* = 0.26$ me, $v \sim 2 \times 10^5$ m/s, and $\tau \sim 100$ fs, the estimate for l is 20 nm. The materials which satisfy this dimensional constraint, the scattering from the interface boundary must play an important role in the calculation of charge-carrier transportation. This argument is based on three important underlying assumptions which are stated as follows, firstly, the carrier velocity should be equated to the Fermi velocity (v_F) rather than be equal to the thermal velocity. The electrons which occupies energy states within k_BT of the Fermi surface will be able to perform excitation from an occupied state to an unoccupied state. When the material is excited through photons of energy very high as compared to the bandgap, then the mean carrier velocity will be a function of time which considers the fact the charge carriers will decay to an equilibrium energy state after the excitation source is removed. Secondly, the terahertz electric field can provide electrons significant amount of energy when E_{THz} is larger compared to the electron affinity. Thirdly, at the scale of tens of nanometers, quantum confinement effects can become significant and could play an important role in the conductivity. For instance, in a GaAs quantum well with infinite energy barriers the spacing between the lowest two electron sub-bands when $L = 63$ nm corresponds to 1 THz. Therefore, it can be concluded that the phenomenological conductivity models are applicable when the average electron displacement is approximately equals to the dimension of device but when the quantum confinement effects are not of significant magnitude.

The conductivity of charge carrier undergoing restricted motion is derived from an extension of the Drude-Lorentz model, in which the charge carrier conserves a part of its initial velocity upon scattering from a boundary layer energy barrier [25]. It is assumed that the collisions are randomly distributed in time and c_n denotes the fraction of initial velocity of charge carrier which is conserved after p scattering events in time. With these numerical assumptions, the Drude-Lorentz model can be generalized into a different mathematical form which can be written as,

$$\sigma(\omega) = \sigma_D(\omega)\left[1 + \sum_{p=1}^{\infty} \frac{c_p}{(1 - i\omega\tau)^p}\right] \tag{17}$$

$\sigma_D(\omega)$ denotes the conductivity of Drude-Lorentz model, and c_p is the expectation value $\langle \cos\theta \rangle$ for scattering angle θ. Statistically, if the scattering is a total random process then the carrier's momentum will be randomized which gives $c_p = 0$, on the other hand if the carrier is completely backscattered then $c_p = -1$. The infinite series of Eq. (17) is generally truncated at $p = 1$, which corresponds to the assumption that the charge carrier conserves a part of its initial momentum after the first scattering event, but in every subsequent scattering events the velocity is randomized. This model is called as Drude-Smith model. The Drude-Smith model can be overlapped with surface plasmon model when $c_1 = -1$ and $c_{p>1} = 0$, and with the following mathematical substitution,

$$\omega_0^2 = \frac{1}{\tau_{DS}^2} \tag{18}$$

$$\frac{1}{\tau_{SP}} = \frac{2}{\tau_{DS}}. \tag{19}$$

The subscripts DS and SP are used to distinguish the scattering times denoted in the Drude-Smith model and surface plasmon model. There is an alternate approach to simulate the conductivity of materials with length scales comparable to the mean free path which includes the effect of non-uniformity on the electronic states. The non-uniformity of the electronic states is the result of disordered materials for which long-range delocalized eigen states such as Bloch functions cannot be used to describe the electronic wavefunctions. Tight-binding approach more appropriately describes the localized states of disordered material [26]. The first order correction to the Drude-Lorentz model take the following form in the limit of weak disorder [26–29],

$$\sigma_1(\omega) = Re\left[\sigma_D(\omega)\right]\left(1 - A\tau^{-2} + (3\omega)^{1/2} A\tau^{-3/2}\right) \tag{20}$$

where A is the fit parameter ($A = C/(k_F v_F)^2$) and it is dependent on the Fermi wavevector (k_F), Fermi velocity (v_F) and a constant C of order unity. This conductivity expression is referred to as the localized-modified Drude-Lorentz model and can be derived from the quantum conductivity model which is discussed later in the chapter.

Drude-Lorentz model is attempted to provide another extension which include the influence of an energy dependent τ. The Cole-Cole and Cole-Davidson models add exponents to the denominator of Eq. (12) producing the following conductivity equation,

$$\sigma(\omega) = \frac{N \times e^2}{m^*} \times \frac{\tau}{\left(1 - (i\omega\tau)^{1-\alpha}\right)^{\beta}} \tag{21}$$

where the Cole-Cole (α) and Cole-Davidson (β) parameter are in the range $0 < \alpha < 1$ and $0 < \beta < 1$. The Drude-Lorentz model is the special case of the model of Eq. (21) where $\alpha \to 1$ and $\beta \to 1$. This expression can model small deviations from the Drude-Lorentz conductivity but does not provide elaborate insight into the dominant scattering mechanisms. Most importantly, it cannot reproduce the negative imaginary conductivities which are reported for semiconductor nanomaterials, and this leads to further development of conductivity models.

3.4 Boltzmann conductivity model

A classical approach has been used to model the terahertz conductivity where quantum mechanical effects such as a finite density of energy states, or an energy dependent τ, has not been accounted for in the mathematical equation. Boltzmann transport equation can be used to derive a generalized expression of conductivity in which both the contributions are addressed quantitatively. Under an external perturbation such as an electric field, it is assumed that the rate of change of the electron distribution function is inversely proportional to the scattering time τ, this is called as relaxation time approximation. The frequency dependent conductivity of an isotropic three-dimensional material with parabolic energy bands is derived from the Boltzmann transport equation which is given by,

$$\sigma(\omega) = \frac{2\sqrt{2m^*}e^2}{3\pi^2\hbar^3}\int_E \frac{\tau(E)E^{3/2}}{1 - i\tau(E)\omega}\left(-\frac{\partial f_0}{\partial E}\right)dE. \tag{22}$$

The electron distribution function in equilibrium at temperature T is given by the Fermi-Dirac distribution,

$$f_0(E) = \frac{1}{e^{(E-\mu)/k_B T} + 1} \tag{23}$$

where the chemical potential μ can be determined mathematically from the given numerical equation,

$$N = \int_0^\infty f_0(E)g(E)\,dE \tag{24}$$

where $g(E)$ denotes the electronic density of states. Fermi's golden rule can be applied to calculate the energy dependent scattering times $\tau(E)$ for various scattering mechanisms such as with impurities, acoustic or optical phonons. This approach has the benefit of converting the scattering time which is a fit parameter into a numerically determined quantity obtained from the carrier density of semiconductor. This model provides a physical insight into the underlying microscopic physics because the calculated scattering time $\tau(E)$ will highlight the dependence on the inclusion of relevant scattering mechanisms. In addition, it can be inferred from the model that the conductivity is dominated by the contribution of electrons with energies close to the chemical potential (μ), this can be understand from the fact that the peak value of the function $\partial f_0/\partial E$ comes at the energy $E = \mu$. The function $\partial f_0/\partial E$ is given by the equation,

$$\frac{\partial f_0(E)}{\partial E} = \frac{e^{(E-\mu)/k_B T}}{k_B T\left(e^{(E-\mu)/k_B T} + 1\right)^2}. \tag{25}$$

1. Skin depth is much less than linear dimensions of metal, thus regarding it as planar and infinite in extent.

2. Normal incidence of propagation, thus simplifying to a one-dimensional problem.

3. Conduction electrons are quasi-free, having a kinetic energy $E(k) = \hbar|k|^2/2m$ with a parabolic band approximation, where $\hbar$ = modified Planck's constant.

4. Collision mechanism is always described in terms of l_m.

5. A fraction p of electrons arriving at the surface is scattered specularly, while the rest are scattered diffusely.

6. This one-band free-electron model does not apply to multivalent metals, in which the electrons occupy more than one energy band (e.g., aluminum or tin).

7. $K(s)$ is derived for transverse conductivity, and holds for the whole (q, ω) plane, for a spherical energy band.

8. While the semiclassical treatment takes into account partially-filled conduction band energy dispersion, $E(k)$, it ignores electron wavefunctions.

These assumptions laid the foundation of a more accurate theoretical framework which considers the quantum behavior of electrons for the purpose of calculating the charge carrier dynamics inside the semiconductor materials.

3.5 Quantum conductivity model

The conductivity associated with interband and intraband transitions can be determined accurately from the linear response theory or Kubo-Greenwood theory. The transition rate of electrons denoted by W_{ij} between an initial state i and a final state j is calculated by the Fermi's golden rule, and $\hbar\omega W_{ij}$ is equated to the power lost in the form of electromagnetic field due to absorption (mathematical formulation of this statement yields the absorption coefficient α). Further, the real part of the refractive index can be determined from the absorption coefficient (α) by the Kramers-Kronig relations. The resulting expression for the conductivity is given by the equation,

$$\sigma(\omega) = \frac{2e^2\hbar}{(m^*)^2V}\sum_{ij}\frac{\left|\left\langle\psi_j|\hat{e}\cdot p|\psi_i\right\rangle\right|^2}{E_j - E_i} \times \frac{i[f(E_j) - f(E_i)]}{E_j - E_i - \hbar\omega - i\hbar/\tau} \tag{26}$$

where $\hat{e}$ denotes the unit vector in the direction of the electric field $(\vec{E})$, p is the momentum operator and V represents the volume in which the summation of the dipole matrix element $\left|\left\langle\psi_j|\hat{e}\cdot p|\psi_i\right\rangle\right|^2$ is applicable. The matrix element creates selection rules for observable absorption peaks, for instance in semiconductor quantum wells intersubband absorption is only observed if the applied electric field is parallel to the direction of quantum confinement, and if the initial and final states have parity. It is important that the wavefunctions (ψ) of initial and final states are already determined, to calculate the quantum expression of conductivity. Additionally, a common formalism should be designed to include the interband, intersubband and intraband transitions. Boltzmann expression of Eq. (22) is the result of solving Eq. (26) for intraband transitions in a parabolic band. The conductivity expression of Eq. (26) can be simplified for intersubband transitions in quantum wells, such as absorption from lowest state 1 to the next highest state 1, this yields the real part of the conductivity which is given as,

$$\sigma_1(\omega) = \frac{e^2 f_{10}}{2m^*} \int g(E) f(E) \frac{\hbar \Gamma(E)}{(\hbar\omega - E_{10})^2 + \Gamma^2(E)} dE, \tag{27}$$

where f_{10} is the oscillator strength and the linewidth in units of energy is denoted by $\Gamma = \hbar/\text{tau}$. The expression can further be solved in the limit of an energy-independent Γ, which yields the following equation,

$$\sigma_1(\omega) \propto \frac{\Gamma}{(\hbar\omega - E_{10})^2 + \Gamma^2}. \tag{28}$$

The most prominent systems to form excitons from the binding of electrons and holes together, are the ones which exhibits quantum confinement and where the excitonic binding energy is enhanced by the quantum effects. The model of hydrogen atom can be used to derive a comparison for the eigen states of an exciton, where 1s, 2s, 2p, 3s, ... are in spectroscopic notation. It is important to highlight the difference between the classical approach and the current approach, for the hydrogen atom, the lowest order electric dipole allowed transition (1s-2p) falls in the ultraviolet range while for GaAs quantum wells due to the lower effective masses of electrons and holes, and high dielectric constant, 1s-2p transition results in the excitons of terahertz frequency range. The expression of frequency dependent complex conductivity is yielded from the Fermi's golden rule,

$$\sigma(\omega) = \frac{N_X e^2}{m^*} \sum_k \frac{f_{j,k}\,\omega}{\omega_{j,k}^2 - \omega^2 - i\omega/\tau_X} \tag{29}$$

where N_X and T_X denote the excitonic density and linewidth, $\hbar\omega_{j,k}$ denote the energy difference between the initial state j and final state k, and $f_{j,k}$ is the electric dipole matrix element for the transition.

Further, for superconducting materials, Mattis-Bardeen theory provides the quantum mechanical expression for the conductivity. When the terahertz photon energy exceeds the Cooper pair binding energy 2Δ, this results in strong absorption and production of a σ_1 that increases with frequency above $\hbar\omega = 2\Delta$. Apart from the Mattis-Bardeen theory, there is another model called as two-fluid model to simulate the conductivity and is often utilized to approximate the conductivity in the terahertz range. The condition associated with the applicability of two-fluid model states that the terahertz photon energy should be very less than the Cooper pair binding energy ($E < < 2\Delta$). In this model, the conductivity is the summation of two terms, first is the Drude-Lorentz term which arises from the thermally excited carriers and the second is the superconducting term (σ_s) which is purely imaginary. The fraction (x_S) of carrier in superconducting state is zero at the superconducting transition temperature while the the fraction (x_n) of thermally excited carrier is zero at zero temperature. The conductivity is given by the following equation,

$$\sigma(\omega) = \frac{Ne^2}{m^*}\left[\frac{x_n\tau}{1-i\omega\tau} + x_s\left(\frac{i}{\omega} + \pi\delta(\omega)\right)\right] \quad (30)$$

where N denotes the sum of the densities of the superconducting Cooper pairs and the normal carriers, and $x_s + x_n = 1$. By the Kramers-Kronig relations, it can be stated that the infinite DC conductivity in the superconducting state is inversely proportional to the frequency ($\sigma_2 \propto 1/\omega$).

The extended Drude model provides an additional accurate approach to simulate the terahertz conductivity in quantum materials [30, 31]. It has been extensively applied to the visible and infrared conductivity of metals and superconductors, and is applicable for Fermi liquids with strong electron–electron interactions, rather than the weaker interaction of a Fermi gas. In this model, the single-particle effective mass (m^*) in Eq. (12) is replaced by a frequency-dependent effective mass ($m^{**}(\omega)$ which is given by the equation,

$$m^{**}(\omega) = m^*[1 + \lambda(\omega)] \quad (31)$$

and the energy-independent scattering time τ is replaced by a photon frequency-dependent function $\tau^*(\omega)$. The effective mass renormalization factor $1 + \lambda(\omega)$ can be experimentally determined from the electron density N by using the equation,

$$m^{**}(\omega) = -Ne^2 \times Im[1/\sigma]/\omega. \quad (32)$$

4. Experimental determination of complex conductivity

Generally, the studies based on broadband spectroscopy are obtained from a method known as terahertz-time domain spectroscopy (THz-TDS) which measures the electric field of pulses of electromagnetic radiation directly in the real time after the interaction of radiation with the sample material. Short pulses (<100 fs) of infrared radiation of centre wavelength 800 nm are used in THz-TDS, for example: mode-locked Ti: sapphire laser can be used to generate and detect single-cycle pulses of the radiation. It is known that the time and frequency resolution are inversely proportional to each other, and as the duration of pulses are in sub-picosecond range, this results in a broadband frequency spectrum with significant amount of power from tens of GHz to a few THz. There are other alternate methods to generate terahertz radiation, firstly the down-conversion of IR radiation into the THz range for which non-linear techniques such as difference frequency generation, optical rectification and gas plasma generation are the efficient methods for pulses of high fluences [21, 32]. Secondly, photoconductive emitters can be utilized to generate a pulse of lower fluences where a transient photocurrent generates a pulse of THz radiation [33, 34]. The photocurrent can be a result of various fields that cause charge separation such as an applied electric field (Auston switches), field due to the surface accumulation layer or depletion layer [35, 36], and photo-Dember fields which are created by the difference in the ballistic motion of electron and hole due to the effective mass [36, 37].

Now, lets discuss about the detection methods, there are generally two competing technologies that are being used for the coherent detection of THz pulses which are as follows,

1. The first detection method is based on the change in the polarization state of an IR gate pulse due to the interaction with the electric field of THz pulse, this

is known as electro-optic sampling [38, 39]. The polarization state of the pulse is often analyzed by a Wollaston prism and balanced photodiodes, and the electric field of THz pulse can be directly calculated from the experimental signals using known parameters of the electro-optic crystal [40].

2. The second method is the photoconductive detection of THz radiation pulses which is basically a inverse process of emission [34]. A photocurrent created by a gate beam flows between two contacts under the influence of the THz electric field, producing a signal that is proportional to E_{THz} (when the photoconductive decay time of the material is very short) [41].

Commonly, a THz-TDS setup contains a photoconductive antenna which generates and detect THz radiation. The emitter chip generates a burst of terahertz radiation when excited by the photons of a subpicosecond laser pulse of average beam power in the range of few mW. A silicon lens is generally included in the setup which collimates the excited THz pulse for parallel free space propagation before reflecting from a parabolic mirror. A second mirror reflects the THz pulse onto the detector, and the time-dependence of the THz pulse is obtained by scanning the relative time delay between the emitter excitation pulse and the detection pulse. The conductive sample which is located between the parabolic mirrors, causes a time delay and a reduction in the amplitude of the THz pulse which can be compared with the reference THz pulse measured without the sample. This comparison gives the information about the magnitude and phase of the spectra and hence results in the numerical calculation of absorption coefficient and refractive index. Hence, the conductivity (σ) of the sample is determined experimentally as the it is defined by the quantity of absorption and refractive index. However, THz-TDS system has a number of advantages including the ability to extract complex optical properties of material, there are of course some disadvantages of the system, such as the high cost of the technology and a narrow frequency range as compared to the Fourier-transform infrared spectrometers. Most importantly, the reflection geometry needs to be precisely monitored to prevent errors entering into the calculated optical properties of the sample [42, 43].

Compared with the THz-TDS method, Time-Resolved Terahertz Spectroscopy (TRTS) is an another important method to probe the conductivity at higher frequencies and is therefore sensitive to THz conductivity instead of static conductivity of materials. The sub-picosecond time resolution is suitable to study the ultrafast dynamics of carriers. The complex photoconductivity is calculated from the real time monitoring of amplitude and phase of emitted THz wave which is excited by the optical pump probe. For THz-TDS, electric field waveforms transmitted through the unexcited sample, $E_0(t)$, and reference, $E_i(t)$ are scanned and averaged which typically results in waveforms with a reproducible measured phase delays of femtoseconds range. However, for TRTS measurements, the pump delay is scanned while holding the delay between the gate and terahertz probe pulses fixed at the position that gives the maximum differential electro-optic response ΔE_{EXC} to yield the TRTS decay dynamics. The optical pump-probe induced change in conductivity can be calculated by collecting and averaging the data from pump delay scans. Spectral changes to the photoconductivity is determined by measuring the differential electrical field waveforms, $\Delta E_{EXC}(t)$, where the delay time between the THz probe and gate is scanned and the delay between the visible pump and THz probe is kept constant. Further, the conductivity of photoexcited samples is determined by analyzing the measured electric field waveforms of $\Delta E(\text{peak})$ under the thin-film approximations. The electric field transmission through a sample consisting of a photoexcited layer smaller than the overall sample thickness relative to an non-photoexcited sample is given by, [44].

$$\frac{\hat{T}_s^p(\omega)}{\hat{T}_s^{np}(\omega)} = \frac{\hat{E}_s^p(\omega)}{\hat{E}_s^{np}(\omega)} = \frac{1 + n_{sub} + Z_0 d^{np} \tilde{\sigma}^{np}(\omega)}{1 + n_{sub} + Z_0 d^p \tilde{\sigma}^p(\omega)} \tag{33}$$

where the substrate is defined as the unexcited portion of the sample and the superscripts np and p indicate the non-photoexcited and photoexcited samples respectively. The rearrangement of the expression in Eq. (33) gives the explicit representation for the conductivity of the photoexcited sample. The real part of the conductivity determined by Eq. (33) is related to the direct current conductivity in the low frequency limit by $\sigma = eN\mu$, where N and μ are the charge carrier density and mobility, respectively, and e is the electron charge. In general, bulk semiconductors generate equal number of both electrons and holes with unit quantum yield following photoexcitation so the conductivity is given by a sum of the electron and hole contributions. This analysis method therefore provides the sum of the electron and hole conductivity in the photoexcited sample regardless of the intrinsic carrier type present and hence can be utilized directly to doped-semiconductor samples.

5. Charge carrier dynamics in hydrogenated amorphous silicon

Solar energy is one of the clean source of technology that has the potential to reduce the anthropogenic damage to the earth and its ecosystem. This puts the photovoltaic materials that make up solar cell as the subject of intensive study and scientists all around the world are analyzing the photovoltaic properties of novel materials to discover the true potential of solar energy [45, 46]. Time Resolved Terahertz Spectroscopy is one of the methods to analyze the conductivity of photovoltaic materials where spectroscopy, in general defines the study of interactions between matter and electromagnetic radiation. The sentence has been corrected. Generally, the field of spectroscopy plays an important role in the study of photovoltaic materials which is necessary for the efficient designing of solar cells. The investigation of early-time film dynamics offers unique opportunities to improve the photoconductive characteristics of the active materials by better understanding the carrier evolution properties.

Terahertz spectroscopy uses ultrashort pulse lasers to study the charge carrier dynamics of matter and electromagnetic radiation within extremely short time scales (nanoseconds to picoseconds). The lasers excite the specimen material in short pulses, which initiates the generation of charge carriers in the materials. Hence, provides a non-contact and accurate method to measure the photo conductivity of the material providing insight into the nature of charge carriers, respective mobility and time dependence of the conductivity. Hydrogenated amorphous silicon (a-Si:H) thin films are attracting increasing attention for their applications to silicon hetrojunction solar cells in surface passivation [47]. The state-of-art deposition method of intrinsic a-Si:H thin films is plasma enhanced chemical vapor deposition (PECVD), in this section we test the passivation capabilities of a-Si:H films on n-type silicon (Si) surface by measuring the change in conductivity ($\Delta\sigma$) of the material after the optical excitation as expressed by the following equation,

$$\Delta\sigma = \xi \times (\mu_e + \mu_h) = -\frac{\Delta E_{exc}(\omega)}{\Delta E_{gs}(\omega)} \times \frac{\varepsilon_0 c}{Fe_0} \times \frac{1}{1 - e^{-\alpha L}} \tag{34}$$

where ξ is quantum yield of charge generation, μ_e and μ_h are the electron and hole mobility, respectively, ΔE_{exc} is the terahertz electric field transmitted through the sample after photo excitation while ΔE_{gs} is the ground state terahertz electric field, ε_0 is permittivity of vacuum, c is velocity of light, F is the fluence in number of photons/cm^2, e_0 is the elementary charge, α is the absorption coefficient, and L the thickness of the sample. The quantity that is obtained from the Eq. (34) has the unit of mobility in cm^{2}V^{-1} s^{-1} which is defined as the product of two quantities namely, quantum yield and mobility. The quantum yield is assumed to be unity ($\xi = 1$) which means that all absorbed photons are converted to mobile charges. The interplay between the time-dependent change in charge population and mobility defines the shape of the terahertz photoconductivity kinetics. On one hand, the rise in the photo conductivity kinetics elucidates generation of charged species and/or increase in mobility of the charges. On the other hand, decay represents decrease of the mobility due to relaxation and/or disappearance of charge carriers by recombination mechanisms.

Intrinsic a-Si:H films are deposited by an oxford instruments PECVD system using a source of silane gas (SiH_4), where the hydrogenation of films is achieved by providing hydrogen gas (H_2) for the generation of plasma in the deposition chamber. The total pressure inside the chamber is maintained around 400 mTorr with the partial pressure of SiH_4 and H_2 in the ratio of 5:4, which is determined experimentally for the optimum hydrogenation of amorphous silicon. Firstly, the samples are RCA cleaned and then, a dry oxidation of the sample is performed to grow a 100 nm double-sided layer of silicon dioxide (SiO_2) on the silicon sample. Secondly, a forming gas annealing at 300°C is performed followed by the conductivity measurement to determine the reference measurements for the comparison purposes. Thirdly, a wet etching procedure using HF is performed to remove the top-layer of SiO_2 and lastly, the sample is immediately placed in the chamber for the deposition of 30 nm of a-Si:H for surface passivation. The thickness of deposited amorphous silicon films are measured using a J.A. Woollam N-2000 spectroscopic ellipsometry system.

The schematic of the experimental setup is shown in **Figure 2** and the results prepared for the measurements performed are shown in **Figure 3**. A pump probe delay time is varied in order to obtain the photoconductivity kinetics while the gating delay time is fixed at the peak of the terahertz electric field (E_{THz}). The pump-probe delay time is scanned within an interval of 1500 s to record the transmitted terahertz electric field. It should be noted that the photoconductivity obtained with this process represents the lower limit of the mobility (**Figure 3(c)**). The transmitted terahertz electric field (E_{GS} and E_{EXC}) responses can be fitted to

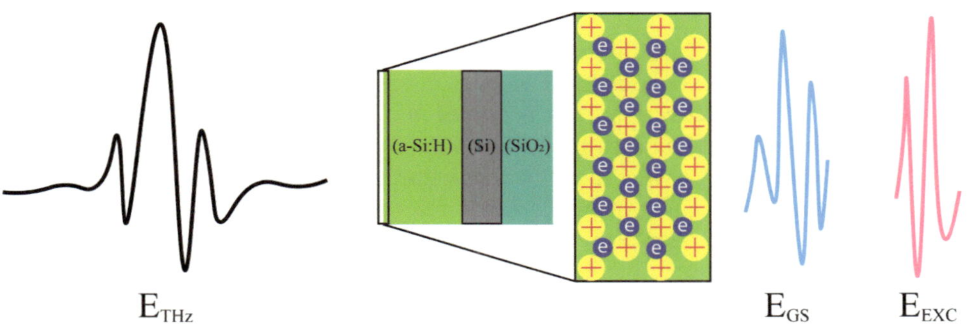

Figure 2.
Schematic diagram of the terahertz setup used in probing charge carrier dynamics in hydrogenated amorphous silicon passivation layer. The transmitted electric fields, E_{GS} (ground state) and E_{EXC} (excited state), are Fourier transformed and used the conductivity Eq. (34) to obtain the photoconductivity kinetics at the maximum of terahertz electric field (E_{THz}).

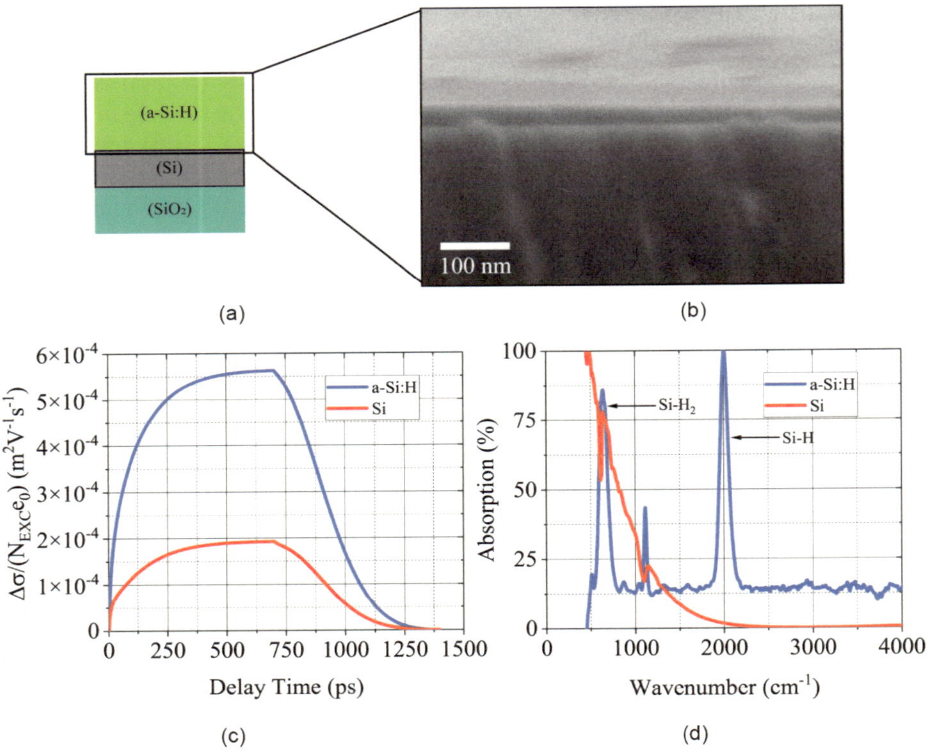

Figure 3.
Surface passivation effectiveness of a layer of hydrogenated amorphous silicon (a-Si:H). (a) Schematic of the double-layered sample under observation, it consists of one-sided growth of 100 nm SiO_2 and on the other side 30 nm a = Si:H is deposited through PECVD process. (b) Scanning Electron microscopy (SEM) image of the top surface of the sample where an uniform layer of a-Si:H is clearly visible due to contrast difference between Si and a-Si:H. (c) the photoconductivity kinetics of the sample under study normalized to the excitation density where $N_{EXC} = 1 \times 10^{15}$ and e_o is the elementary electronic charge. The effectiveness of surface passivation from a-Si:H can be stated from the difference in the terahertz conductivity between the passivated Si sample and non-passivated Si sample. (d) the results of Fourier transform infrared spectroscopy (FTIR) of the a-Si:H sample compared with the Si substrate to highlight the presence of Si − H and Si − H_2 bonds in the passivation layer.

Drude or Drude-Smith model to elucidate the change generation efficiency (quantum yield) and the mobility independently. The results of the photoconductivty point to a very important parameter that controls charge dynamics in photovoltaic materials: defects and dangling bonds, as evident by the difference in the photoconductivity kinetics of a-Si:H-passivated silicon sample and non-passivated silicon sample (**Figure 3(c)**). Through the use of time resolved terahertz spectroscopy, these early time processes and parameters of novel solar cell materials can be investigated in nanometer scale. The presence of hydrogen (H) in the passivation layer is by using Fourier Transform Infrared Spectroscopy (FTIR) and the many repetitive measurements were taken to check that no hydrogen loss takes place with time. The integration times were sufficient so as to reduce the statistical errors to a few percent of the average value. Infrared absorption measurements were made with a Perkin Elmer spectrometer on films deposited on crystalline silicon. The infrared spectrum of hydrogenated a-Si generally displays two groups of bands due to the stoichiometric ratio of silicon and hydrogen i.e. Si − H_X bonds. The first group is designated to the bond wagging (rocking, rolling) band at $\approx$640 cm^{-1} which is directly proportional to the hydrogen concentrations and the second group corresponds to the band in the region of wavenumber 2000 cm^{-1} where the vibrations are designated by the stretching of the Si-H bonds.

6. Conclusions

Terahertz radiation, in particular the theoretical framework, has become a requirement to design faster and energy efficient devices. Successful first-pass design of components of a system requires the ability to accurately predict the dissipative losses due to charge-carrier scattering in semi-conductor and metallic materials used in waveguides, filters, antennas and optics. A comprehensive description of different formalisms was presented that allows the complex conductivity of semiconductor materials to be calculated at terahertz frequencies by including the energy dependence of the scattering rate. The generalized conductivity reduces to the Drude model in the limit of zero temperature or an energy independent scattering rate. Terahertz time-domain spectroscopy was used to determine the conductivity and the photoconductivity of hydrogen-doped amorphous silicon, which was semi-quantitatively accounted for by the generalized conductivity model. The breadth and depth of the studies of conductive solid state materials using terahertz radiation continues to advance and as the field is very broad, it is very important to understand the fundamental models to better utilized materials for future. Terahertz based spectroscopy methods can be valuable to not only in revealing fundamental processes or micro/nano strategies but also in providing insights for possible optimization of manufacturing procedures for solar cell industry.

Acknowledgements

Rahul Goyal would like to thank Ministry of Human Resource and Development, Government of India, for providing financial support and Centre for Nanoscience and Engineering for the research facilities.

Conflict of interest

The authors declare no conflict of interest.

Author details

Rahul Goyal[1]* and Akash Tiwari[2]

1 Indian Institute of Science, Bangalore, India

2 Indian Institute of Information Technology, Jabalpur, India

*Address all correspondence to: rahulgoyal@iisc.ac.in

References

[1] Siegel PH. Terahertz technology. IEEE Transactions on Microwave Theory and Techniques. 2002;**50**(3): 910-928 Available from: http://ieee xplore.ieee.org/document/989974/

[2] Tonouchi M. Cutting-edge terahertz technology. Nature Photonics. 2007; **1**(2):97-105. DOI: 10.1038/ nphoton.2007.3

[3] Rubens H, Nichols EF. Heat rays of great wave length. Physical Review (Series I). 1897;**4**(4):314

[4] Mittleman DM. Perspective: Terahertz science and technology. Journal of Applied Physics. 2017; **122**(23):230901

[5] Vijayraghavan K, Jiang Y, Jang M, Jiang A, Choutagunta K, Vizbaras A, et al. Broadly tunable terahertz generation in mid-infrared quantum cascade lasers. Nature Communications. 2013;**4**(1):1-7

[6] Liu L, Xu H, Percy RR, Herald DL, Lichtenberger AW, Hesler JL, et al. Development of integrated terahertz broadband detectors utilizing superconducting hot-electron bolometers. IEEE Transactions On Applied Superconductivity. 2009;**19**(3): 282-286

[7] Peng K, Parkinson P, Fu L, Gao Q, Jiang N, Guo YN, et al. Single nanowire photoconductive terahertz detectors. Nano Letters. 2015;**15**(1):206-210

[8] Gu J, Singh R, Liu X, Zhang X, Ma Y, Zhang S, et al. Active control of electromagnetically induced transparency analogue in terahertz metamaterials. Nature Communications;**2012, 3**(1):1-6

[9] Grady NK, Heyes JE, Chowdhury DR, Zeng Y, Reiten MT, Azad AK, et al. Terahertz metamaterials for linear polarization conversion and anomalous refraction. Science. 2013; **340**(6138):1304-1307

[10] Tanoto H, Teng J, Wu Q, Sun M, Chen Z, Maier S, et al. Greatly enhanced continuous-wave terahertz emission by nano-electrodes in a photoconductive photomixer. Nature Photonics. 2012; **6**(2):121-126

[11] Hu Z, Kaynak M, Han R. High-power radiation at 1 THz in silicon: A fully scalable array using a multi-functional radiating mesh structure. IEEE Journal of Solid-State Circuits. 2018;**53**(5):1313-1327 Available from: https://ieeexplore.ieee.org/document/ 8254345/

[12] Greene BI, Federici JF, Dykaar DR, Levi AFJ, Pfeiffer L. Picosecond pump and probe spectroscopy utilizing freely propagating terahertz radiation. Optics Letters. 1991;**16**(1):48. DOI: 10.1364/ OL.16.000048

[13] Nuss MC, Auston DH, Capasso F. Direct subpicosecond measurement of carrier mobility of photoexcited electrons in gallium arsenide. Physical Review Letters. 1987;**58**(22):2355-2358. DOI: 10.1103/PhysRevLett.58.2355

[14] Ulbricht R, Hendry E, Shan J, Heinz TF, Bonn M. Carrier dynamics in semiconductors studied with time-resolved terahertz spectroscopy. Reviews of Modern Physics. 2011;**83**(2): 543-586. DOI: 10.1103/RevMod Phys.83.543

[15] Němec H, Kužela P, Sundström V. Charge transport in nanostructured materials for solar energy conversion studied by time-resolved terahertz spectroscopy. Journal of Photochemistry and Photobiology A: Chemistry. 2010; **215**(2–3):123-139. DOI: 10.1016/j. jphotochem.2010.08.006

[16] Adam AJL. Review of near-field terahertz measurement methods and their applications. Journal of Infrared, Millimeter, and Terahertz Waves. 2011; **32**(8):976-1019

[17] Maestrini A, Mehdi I, Siles JV, Lin R, Lee C, Chattopadhyay G, et al. Frequency Tunable Electronic Sources Working at Room Temperature in the 1 to 3 THz Band. Bellingham, WA: Society of Photo-optical Instrumentation Engineers (SPIE); 2012. p. 84960F Available from: http://proceedings.spiedigitallibrary.org/proceeding.aspx?doi=10.1117/12.929654

[18] Mehdi I, Siles JV, Lee C, Schlecht E. THz diode technology: Status, prospects, and applications. Proceedings of the IEEE. 2017;**105**(6):990-1007 Available from: http://ieeexplore.ieee.org/document/7835091/

[19] Ward JS, Chattopadhyay G, Gill J, Javadi H, Lee C, Lin R, et al. Tunable broadband frequency-multiplied terahertz sources. In: 2008 33rd International Conference on Infrared, Millimeter and Terahertz Waves. Pasadena, CA: California Institute of Technology; 2008. pp. 1-3 Available from: http://ieeexplore.ieee.org/document/4665437/

[20] Porterfield DW. High-efficiency terahertz frequency triplers. In: 2007 IEEE/MTT-S International Microwave Symposium. Honolulu, HI, USA: IEEE; 2007. pp. 337-340 Available from: http://ieeexplore.ieee.org/document/4263816/

[21] Löffler T, Kreß M, Thomson M, Hahn T, Hasegawa N, Roskos HG. Comparative performance of terahertz emitters in amplifier-laser-based systems. Semiconductor Science and Technology. 2005;**20**(7):S134-S141. DOI: 10.1088/0268-1242/20/7/003 Available from: https://iopscience.iop.org/article/10.1088/0268-1242/20/7/003

[22] Nienhuys HK, Sundström V. Influence of plasmons on terahertz conductivity measurements. Applied Physics Letters. 2005;**87**(1):1-3. DOI: 10.1063/1.1977213

[23] Parkinson P, Lloyd-Hughes J, Gao Q, Tan HH, Jagadish C, Johnston MB, et al. Transient terahertz conductivity of GaAs nanowires. Nano Letters. 2007;**7**(7):2162-2165. DOI: 10.1021/nl071162x

[24] Pitarke JM, Silkin VM, Chulkov EV, Echenique PM. Theory of surface plasmons and surface-plasmon polaritons. Reports on Progress in Physics. 2007;**70**(1):1-87. DOI: 10.1088/0034-4885/70/1/R01

[25] Smith NV. Classical generalization of the Drude formula for the optical conductivity. Physical Review B—Condensed Matter and Materials Physics. 2001;**64**(15):1-6. DOI: 10.1103/PhysRevB.64.155106

[26] Mott NF, Kaveh M. Metal-insulator transitions in non-crystalline systems. Advances in Physics. 1985;**34**(3): 329-401. DOI: 10.1080/00018738500101771

[27] Jeon TI, Kim KJ, Kang C, Oh SJ, Son JH, An KH, et al. Terahertz conductivity of anisotropic single walled carbon nanotube films. Applied Physics Letters. 2002;**80**(18):3403-3405. DOI: 10.1063/1.1476713

[28] Lee K, Heeger AJ, Cao Y. Reflectance of polyaniline protonated with camphor sulfonic acid: Disordered metal on the metal-insulator boundary. Physical Review B. 1993;**48**(20): 14884-14891. DOI: 10.1103/PhysRevB.48.14884

[29] Jeon TI, Grischkowsky D, Mukherjee AK, Menon R. Electrical and optical characterization of conducting poly-3-methylthiophene film by THz time-domain spectroscopy. Applied Physics Letters. 2001;**79**(25):4142-4144. DOI: 10.1063/1.1427754

[30] Götze W, Wölfle P. Homogeneous dynamical conductivity of simple metals. Physical Review B. 1972;**6**(4):1226-1238. DOI: 10.1103/PhysRevB.6.1226

[31] Basov ND, Timusk T. Electrodynamics of high-superconductors. Reviews of Modern Physics. 2005;**77**(2):721-729. DOI: 10.1103/RevModPhys.77.721

[32] Thomson MD, Kreß M, Löffler T, Roskos HG. Broadband THz emission from gas plasmas induced by femtosecond optical pulses: From fundamentals to applications. Laser and Photonics Reviews. 2007;**1**(4):349-368. DOI: 10.1002/lpor.200710025

[33] Auston DH. Picosecond optoelectronic switching and gating in silicon. Applied Physics Letters. 1975; **26**(3):101-103. DOI: 10.1063/1.88079

[34] Auston DH, Cheung KP, Smith PR. Picosecond photoconducting Hertzian dipoles. Applied Physics Letters. 1984;**45**(3):284-286. DOI: 10.1063/1.95174

[35] Zhang XC, Hu BB, Darrow JT, Auston DH. Generation of femtosecond electromagnetic pulses from semiconductor surfaces. Applied Physics Letters. 1990;**56**(11):1011-1013. DOI: 10.1063/1.102601

[36] Johnston MB, Whittaker DM, Corchia A, Davies AG, Linfield EH. Simulation of terahertz generation at semiconductor surfaces. Physical Review B—Condensed Matter and Materials Physics. 2002;**65**(16):1-8. DOI: 10.1103/PhysRevB.65.165301

[37] Klatt G, Hilser F, Qiao W, Beck M, Gebs R, Bartels A, Huska K, Lemmer U, Bastian G, Johnston MB, Fischer M, Faist J, Dekorsy T. Terahertz emission from lateral photo-Dember currents. Optics Express (2010);**18**:4939-4947

[38] Auston DH, Cheung KP. Coherent time-domain far-infrared spectroscopy. Journal of the Optical Society of America B. 1985;**2**(4):606. DOI: 10.1364/JOSAB.2.000606

[39] Planken PCM, Nienhuys HK, Bakker HJ, Wenckebach T. Measurement and calculation of the orientation dependence of terahertz pulse detection in ZnTe. Journal of the Optical Society of America B. 2001; **18**(3):313. DOI: 10.1364/JOSAB.18.000313

[40] Zhao G, Schouten RN, Van Der Valk N, Wenckebach WT, Planken PCM. Design and performance of a THz emission and detection setup based on a semi-insulating GaAs emitter. Review of Scientific Instruments. 2002;**73**(4):1715-1719. DOI: 10.1063/1.1459095

[41] Castro-Camus E, Fu L, Lloyd-Hughes J, Tan HH, Jagadish C, Johnston MB. Photoconductive response correction for detectors of terahertz radiation. Journal of Applied Physics. 2008;**104**(5). DOI: 10.1063/1.2969035

[42] Nashima S, Morikawa O, Takata K, Hangyo M. Temperature dependence of optical and electronic properties of moderately doped silicon at terahertz frequencies. Journal of Applied Physics. 2001;**90**(2):837-842. DOI: 10.1063/1.1376673

[43] Pashkin A, Kempa M, Nêmec H, Kadlec F, Kužel P. Phase-sensitive time-domain terahertz reflection spectroscopy. Review of Scientific Instruments. 2003;**74**(11):4711-4717. DOI: 10.1063/1.1614878

[44] Jepsen PU, Fischer BM. Dynamic range in terahertz time-domain transmission and reflection spectroscopy. Optics Letters. 2005; **30**(1):29. DOI: 10.1364/OL.30.000029

[45] De Wolf S, Descoeudres A, Holman ZC, Ballif C. High-efficiency

silicon heterojunction solar cells: A review. Green. 2012;**2**(1):7–24

[46] Descoeudres A, Holman ZC, Barraud L, Morel S, De Wolf S, Ballif C. ¿ 21% efficient silicon heterojunction solar cells on n-and p-type wafers compared. IEEE Journal of Photovoltaics. 2012;**3**(1):83-89

[47] Bullock J, Yan D, Wan Y, Cuevas A, Demaurex B, Hessler-Wyser A, et al. Amorphous silicon passivated contacts for diffused junction silicon solar cells. Journal of Applied Physics. 2014; **115**(16):163703